Times of an
Automation
Professional

An Illustrated Guide

The Life and Times of an Automation Professional
An Illustrated Guide

by Ted Williams, Stan Weiner & Greg McMillan

Copyright © 2006   ISA – The Instrumentation, Systems and
Automation Society
67 Alexander Drive
P.O. Box 12277
Research Triangle Park, NC 27709

All rights reserved.

Printed in the United States of America.
10 9 8 7 6 5 4 3

ISBN   1-55617-957-X

No part of this work may be reproduced, stored in a retrieval system, or transmitted in any form or by any means, electronic, mechanical, photocopying, recording or otherwise, without the prior written permission of the publisher.

Library of Congress Cataloging-in-Publication Data in process

Notice

The information presented in this publication is for the general education of the reader. Because neither the author nor the publisher have any control over the use of the information by the reader, both the author and the publisher disclaim any and all liability of any kind arising out of such use. The reader is expected to exercise sound professional judgment in using any of the information presented in a particular application.

Additionally, neither the author nor the publisher have investigated or considered the affect of any patents on the ability of the reader to use any of the information in a particular application. The reader is responsible for reviewing any possible patents that may affect any particular use of the information presented.

Any references to commercial products in the work are cited as examples only. Neither the author nor the publisher endorse any referenced commercial product. Any trademarks or tradenames referenced belong to the respective owner of the mark or name. Neither the author nor the publisher make any representation regarding the availability of any referenced commercial product at any time. The manufacturer's instructions on use of any commercial product must be followed at all times, even if in conflict with the information in this publication.

Reviews

"The best thing about this book is that, if you can't read, you still have a chance of not understanding what they are trying to say."
~ TRUE MEANINGS FROM BEHIND…

"This book could not have been written/drawn 7000 years ago."
~ The Society for Realizing How Far Partial Differential Equations Has Taken Us

"Classic Stan and Greg"   ~ Stan and Greg

"Ted, Greg, and Stan are entering a new dimension in their Quest for the True Meaning of Why We Automate."
~ QTMWWA, September 2005 Edition, QTMWWA@tedgregstan.org

"A must read for anyone who cherishes these adorable automators."
~ AAA (Association for Adorable Automators)

"Industrial-strength cartoons at their finest."
~ (IHOP) Industrial Honorary Ogling Professionals

"This book should not be read by people over the age of 75. I laughed so hard I fell out of my wheelchair and couldn't get up."
~ RAP (Retired Automation Professionals)

"The complete guide to the love and care of automation professionals."
~ SPCA (Society for the Prevention of Cruelty to Automators)

"Ted, Stan, and Greg deserve to be honorary Texans."
~ TEJAS (Texas Exceptional Joint Automation Society)

"The mother of all automation cartoon books."
~ MAD (Mothers of Automation Design)

"Autographed copies to your boss are a great ice breaker for performance reviews."
~ DAP (Dapper Automation Professionals)

"Hoo woo, hoo haa." ~ anonymous famous talk show host

In our hopes of being fair to everyone, we included this paragraph from The Society of Electrical Engineers that Became Options Traders:

"These lunatics will do almost anything to make you believe that controlling variables is more important than good food, sex, or money. They stay at work until all hours of the night in order to convince management that what they do results in more and better products. As anyone with a slightly improved financial knowledge can understand, it is the people in the trading pits that are causing the creation of wealth far in excess of what was deemed sufficient before the advent of the Distributed Control System. We will have more to say about this in future editions."

Acknowledgment

Stan and Greg wish to acknowledge the many funny people that we have worked with in our careers. The conversations at lunch, on the baseball field, basketball court, and during trips were a trip. In particular, we thank John Berra (who left Monsanto to become president in a major automation company without a Muzak system), Bob Cameron (who went to seek his fortune in plastics), Jim Jordan (who left management to become the most insightfully funny president of an engineering firm), Glenn Mertz (who left process control to become a real-estate magnate), and Steve Sanders (who left Monsanto to become rich and famous in electrolyte modeling). Conversations after a few beers with Glenn Mertz were the source of the best top ten lists, and the startlingly unfiltered stream of consciousness from Steve Sanders made him the Robin Williams of Engineering Technology. We also wish to thank the contributors to our "Control Talk" columns and all the technicians, operators, and engineers who would talk to us and whose names we don't remember, thanks to our advanced age. Stan feels that his greatest inspiration came from the managers and vice presidents, who offered such timely and creative solutions to all the control problems.

We also remember with a smile the fun times around the computer screens while camped out in control rooms with Bob Cameron, Henry Chien, Paul Luebbers, Gerald Mitchell, Bob Otto, Mark Sowell, and Terry Tolliver.

The ideas for these cartoons are the culmination of these experiences.

...SAY WHUT?
"...TED, WHAT IS YOUR BATCH RECIPE AND TEMPERATURE CONTROL STRATEGY?"
"...CAN WE AUTOMATE IT FOR ISA-88 BATCH CONTROL STANDARDS?"

"SO... YOU'RE TED WILLIAMS... I THOUGHT YOU WERE IN CRYOGENIC STORAGE..."
"I WAS DEFROSTED TO DO THIS BOOK... GOT ANY ASPIRIN?"
DID YOU PLAY BASEBALL? CAN I HAVE YOUR AUTOGRAPH?

Foreword

Stan and Greg first teamed up with Ted Williams for the ground-breaking book, *How to Become an Instrument Engineer – The Making of a Prima Donna*. Since then, Ted has illustrated numerous sequels plus our monthly "Control Talk" column. Twenty-years later, Ted's cartoons are as fresh, timely, and entertaining as ever.

Stan and Greg spent a total of more than 50 years in the employment of Monsanto so, through association, many readers wonder if Ted Williams is the same person who helped pioneer advances in control theory during the 1960s and then moved on to head a special department and workshop on control at Purdue University and was elected into the "Control Hall of Fame" in 2004.

Others have speculated that Ted is the incredible Boston Red Sox hitter of the 1940s and 50s who is now in the Baseball Hall of Fame. Well, Ted the illustrator is not really related to any of these famous people, but he deserves to be in an industrial Cartoon Hall of Fame. Since we are pretty sure there isn't such a thing, we hereby create one and make *our* Ted the founding member.

To get your friends' and neighbors' attention and to promote the sale of this book, ask them if they have seen the latest from Ted Williams. Some ISA old timers may wonder if the Ted Williams of Purdue is getting out of his rocker in Indiana, but most will think you are off your rocker for proposing the possibility that a cryogenically stored Red Sox player is being thawed out.

The automation profession at times is strange and difficult to explain to people outside the field. You may also remember from our previous writings that it has been extremely difficult explaining it to our managers. It is one of the most misunderstood, yet fascinating jobs in industry. As a part-time professor teaching process control, Greg knows that students in the classroom have no concept of what it really means to be an automation engineer. Therefore, in order to prepare students for industry, we suggest that this book be a required text for every course in control theory.

Automation engineers have some characteristics to match their interesting jobs. This cartoon book shows how you can spot these strange creatures at an early age. Since they are an endangered species, be careful in how you approach them and please don't mess with their habitat. The best place to find them is at ISA conferences where they can be lured for a closer look by offering new software, although the local draft beer has been reported to work as well. Outside of ISA and computer rooms, they are known to spend their time measuring and plotting the equilibrium temperature and declining level of a glass—once filled with ice and ethanol. In Naples, Florida, Stan has made technological advances in reducing the effects of humidity by using insulated glasses.

Back by popular demand are quotes from a control room deep in America's heartland courtesy of Pravin Patel and Kevin Witter. These "strange but true" quotes of operators from the Muscatine plant that first appeared in the book *Dispersing Heat Through Conviction* have been appreciated by people from all walks (and jogs) of life.

The next time friends and relatives ask "what do you do," give them a copy. If the smiles develop into uncontrollable laughter, use your automation skills to prevent injuries from side-splitting humor and, as a last resort, force the reading of standards as an effective antidote.

To help preserve and promote this valuable profession, Greg promises to donate every dollar of royalty that he gets from this book to the profession. At the suspected rate of a quarter per copy, he should have enough to buy a refurbished laptop in 10 years, assuming that the price comes down.

As a Supervisor of the Collier County Soil & Water District in Southwest Florida, Stan will use his share of the royalties to supplement the $9,000,000,000.00 that has been pledged by our governments to restore the Everglades.

A picture is worth a thousand words or, in the case of Stan and Greg, at least two thousand words. This is our chance to sit back and let the cartoons speak for themselves. Enjoy.

"TED... WE NEED CARTOONS ABOUT DISCRETE MANUFACTURING!"
"WHAT IS DISCRETE MANUFACTURING? WHEN I MANUFACTURE SOME-THING, I WANT EVERYONE TO KNOW ALL ABOUT IT!"
SUSAN COLWELL
knock, please
MEN!

NO CHECKS ACCEPTED...
...WAHH!!! MAMA!!!
UH... CAN I HOLD MY BABY?
MOTHER OF INVENTION

From the control room...

"My goats had a set of triplets and a set of doublets." – Wayne

"Is that the tree legged kid?" – Dirck

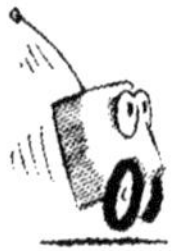

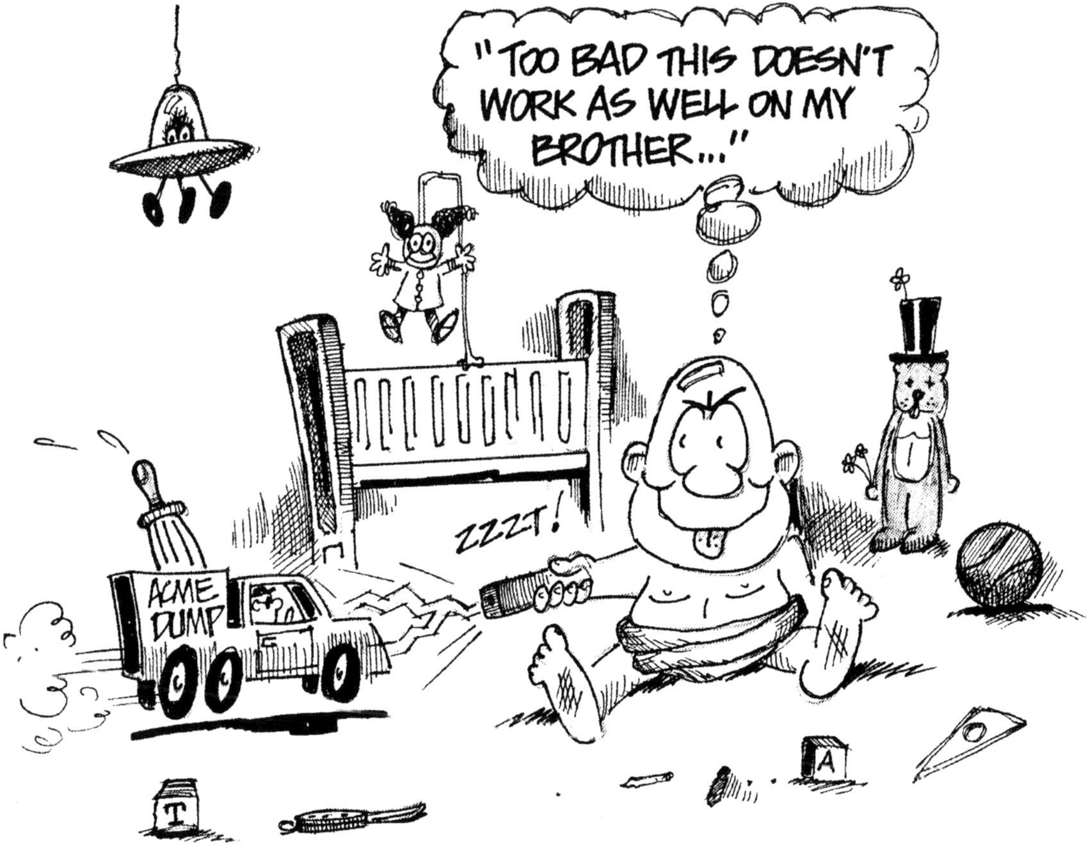
"TOO BAD THIS DOESN'T WORK AS WELL ON MY BROTHER..."
ZZZT!
ACME DUMP
A
T

From the control room…

"Hambone, you are going to be buying diapers and formalin." – Uzi

"Little girls are supposed to be little." – Dub

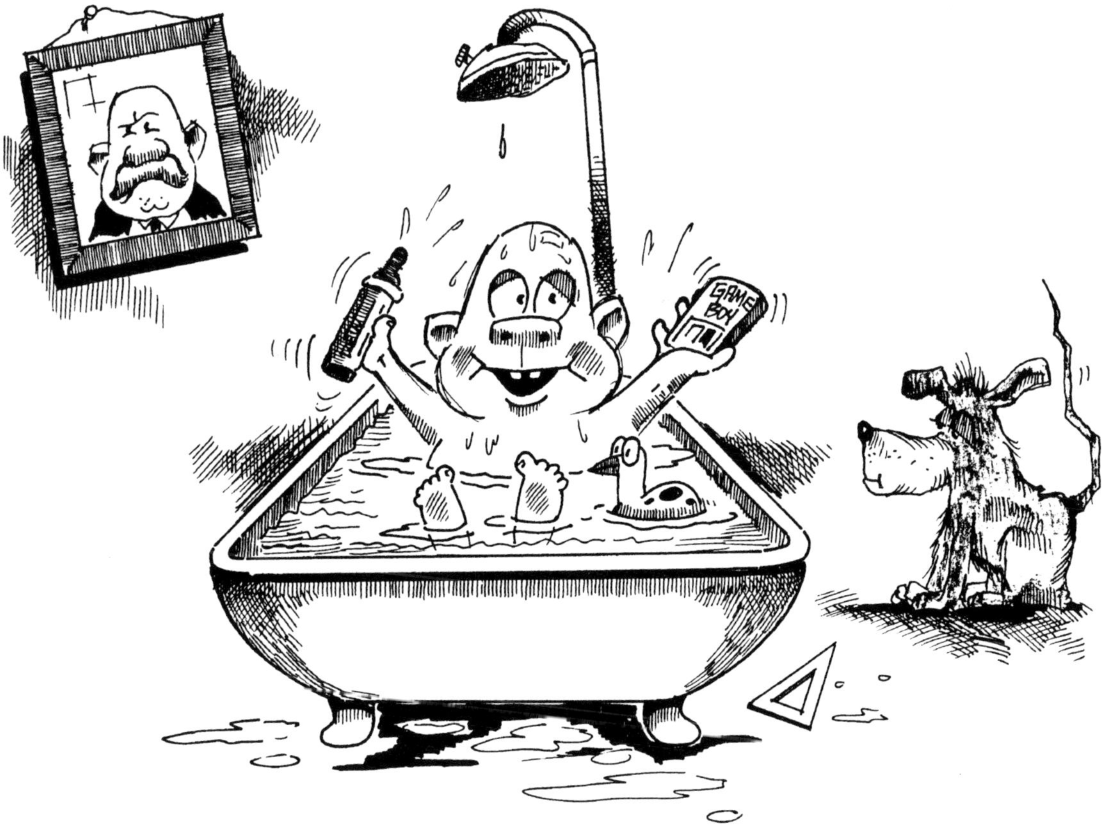
GAME BOY

From the control room...

"I am a bronze goddess." – Jason

"I'm at my ends wit." – Brian M.

...WELL HELEN, AGAIN I COME HOME TO FIND THAT EVERY BUTTON BRINGS UP NICKELODEON... CAN YOU EXPLAIN THAT?... HELEN?

From the control room...

"This procedure reads like stereo instructions." – Mike, after going over a new ACE document

"What's the difference between CDs and BVDs?" – Dub

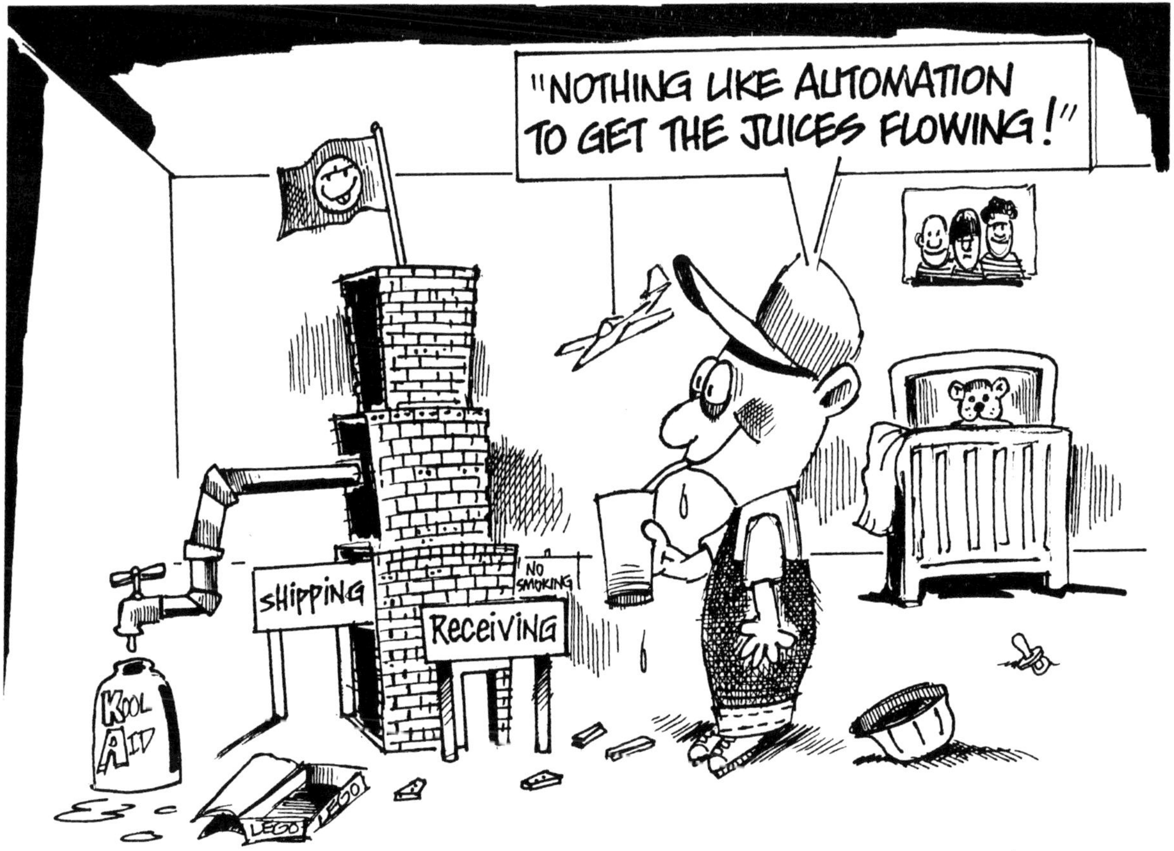
"NOTHING LIKE AUTOMATION
TO GET THE JUICES FLOWING!"
SHIPPING
RECEIVING
NO SMOKING
KOOL AID
LEGO
LEGO

From the control room...

"Do you do this to everyone, or do you do it to other people too?" – Brooke

"You will be someone we read about on TV." – Hambone

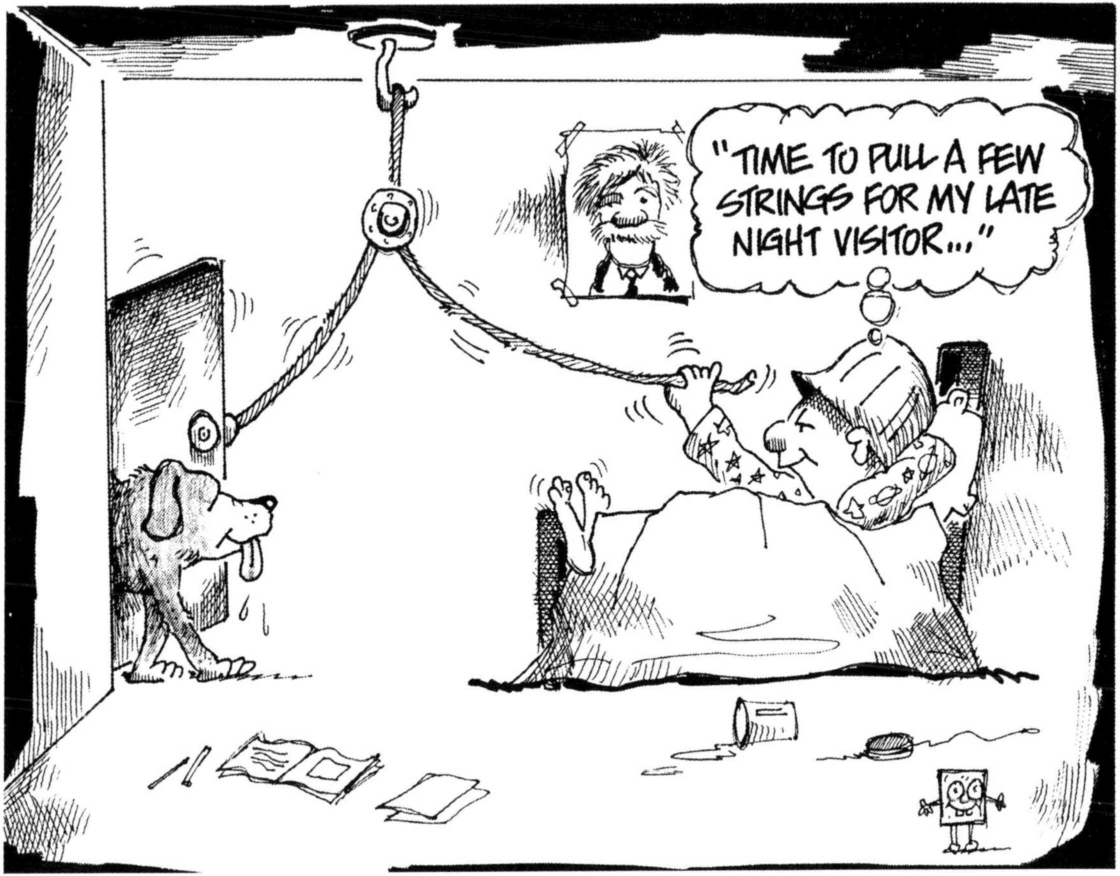
"TIME TO PULL A FEW STRINGS FOR MY LATE NIGHT VISITOR..."

From the control room...

"They got a Wisconsin boy out of Wisconsin." – Ward

"… the right wrench for the tool." – Kevin C.

GOOD BOY!
HMMMMMMMM
NOT... YET... NOT... YET...
I.... WILL... LET YOU... KNOW... WHEN...
NOT.... YET.....
US

From the control room...

"It could be a sheep in skins clothing." – Dodie

"Do you use a dead rabbit call for those coyotes?" – Ward

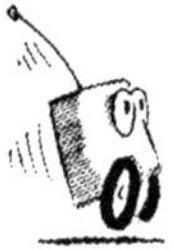

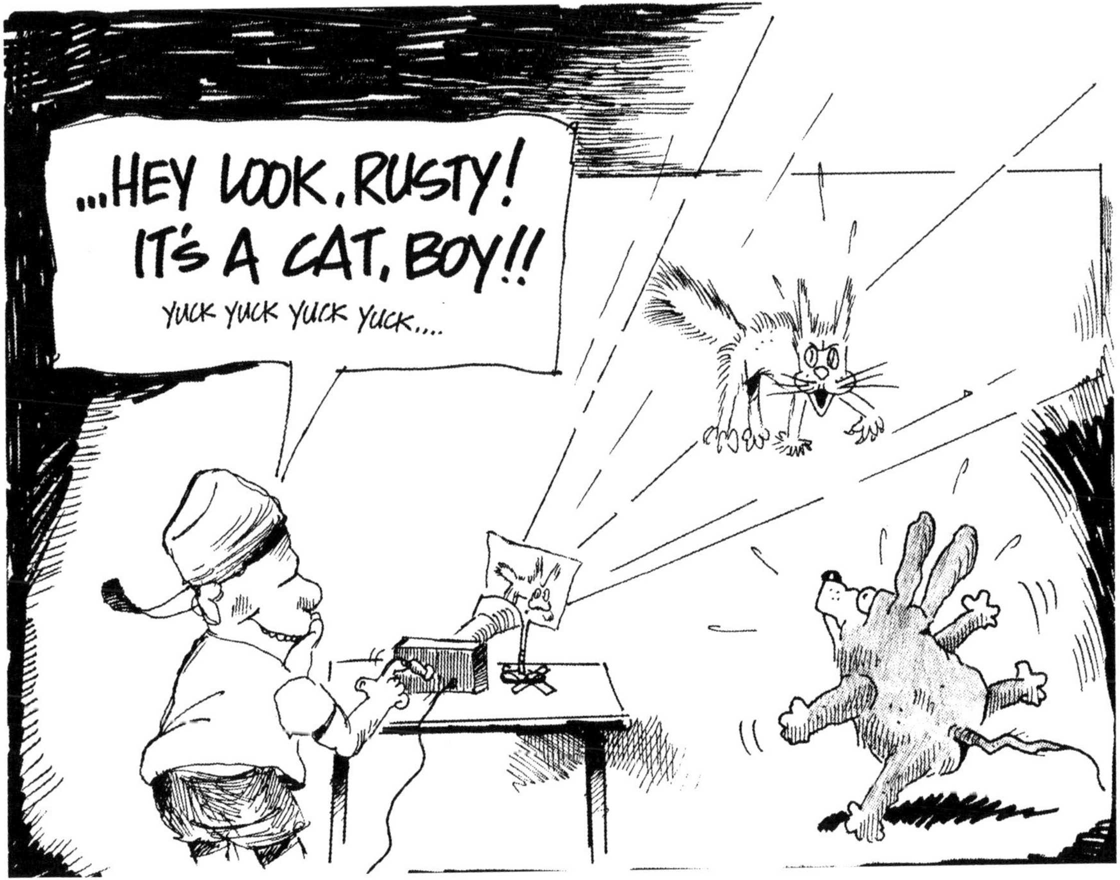
...HEY LOOK, RUSTY! IT'S A CAT, BOY!!
YUCK YUCK YUCK YUCK....

From the control room...

"Who is Halliburton? Is that a fish?" – Thad

"Harkin is in Washington D.C. fifty out of fifty two months a year." – Bill Y.

...YOU WILL NOW RESPOND TO MY COMMANDS... AND MY COMMANDS ONLY!
...the GOVERNMENT!

From the control room...

"My sister is a mailman." – Janet

"All the safety showers out there have water in them." – Brian F.

"WHUT THE..." DOOOHH!!!
THIS'LL HELP KEEP HIM COOL UNDER THE COLLAR... HEH HEH!
US MAIL
9:43

From the control room…

"Around here you have to act like you like people." – Thad

"It's hard to tell what a valve is unless you know what it is." – Dodie

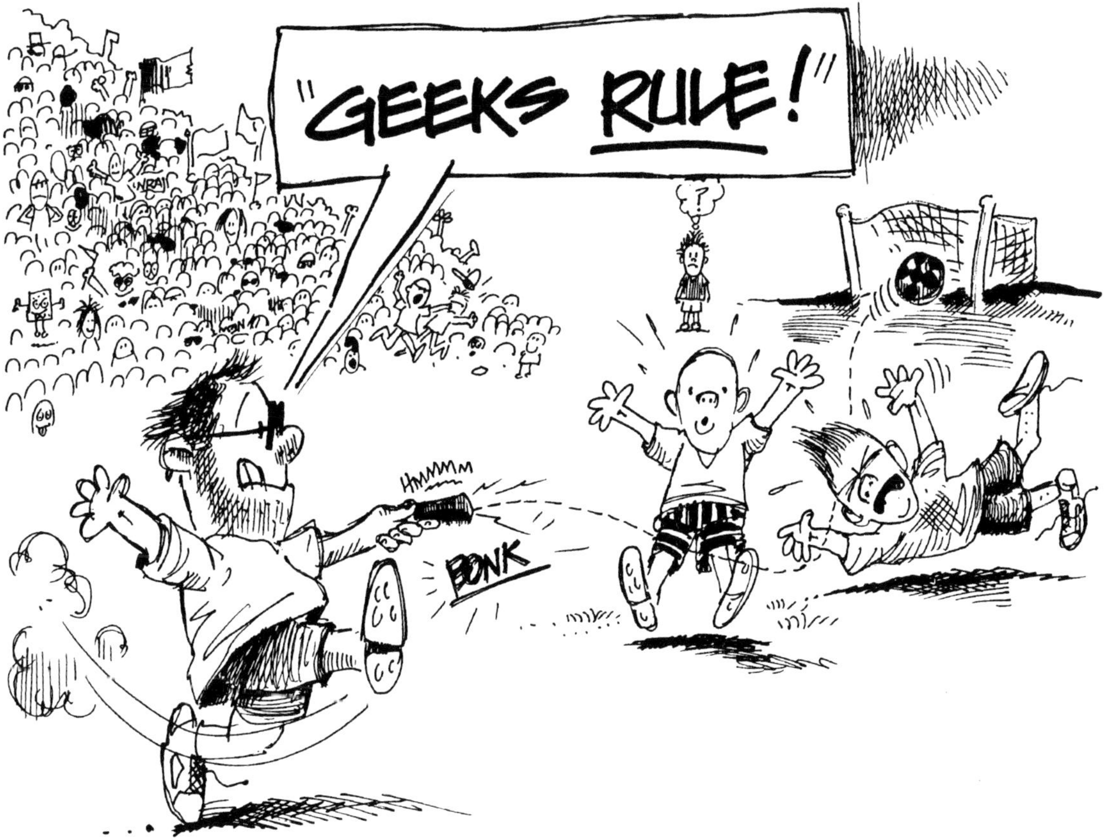
"GEEKS RULE!"
HMMMM
BONK

From the control room...

"Some people sleep with their eyes closed." – Dodie

"45 minutes is way too much time with myself." – Mike M.

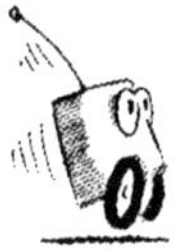

STANLEY! YOU BETTER NOT BE IN THERE READING WITH A FLASHLIGHT, MISTER!
YES, MOM!

From the control room…

"I am somebody. Up til now I wasn't sure." – Jim

"I'm the Drew Carey of the A-Unit." – Ryan

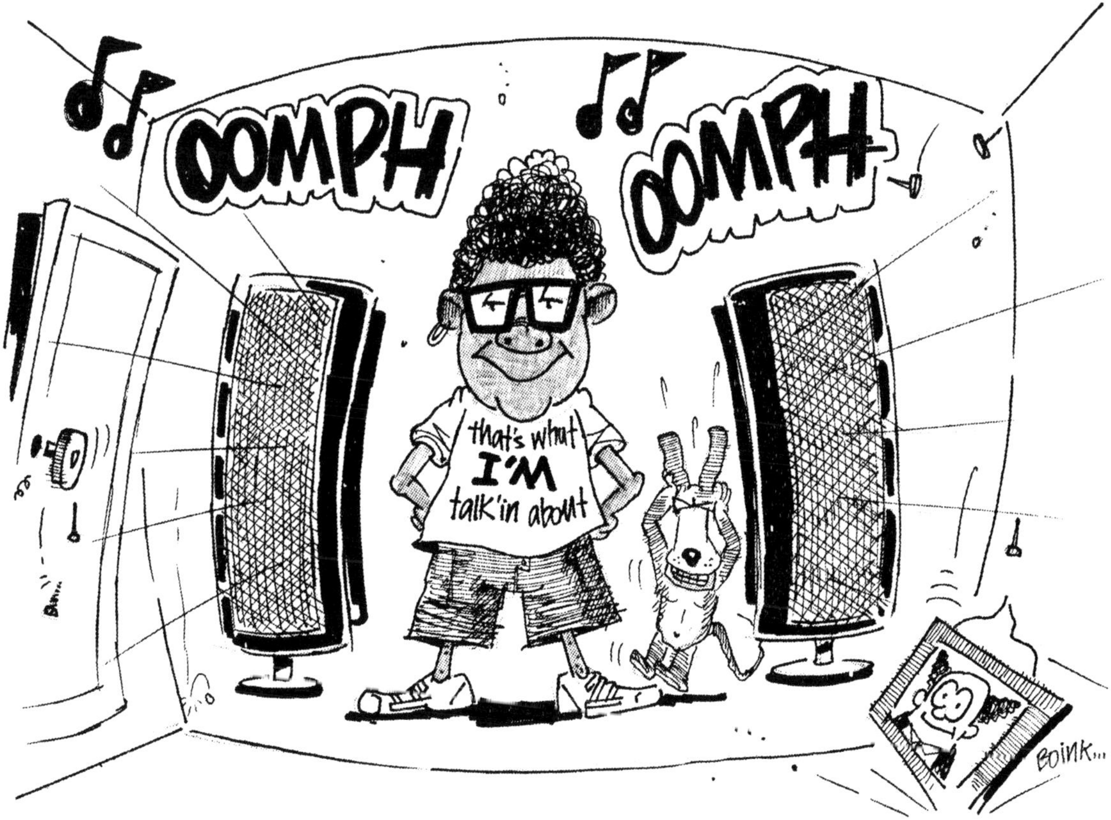
OOMPH
OOMPH
that's what
I'M
talk'in about
BOINK...

From the control room…

"I was never there and I was never caught." – Radar

"Robbie can't even bread his toast." – Quinton

...DID YOU REMEMBER MY LUNCH?
YES, MASTER... (I WISH I WAS STILL A GARBAGE CAN)
the TRUTH IS SOMEPLAC

From the control room...

"Lance Armstrong was the first guy to walk on the moon." – Dub

"How did the Pentagon get its name?" – Jim

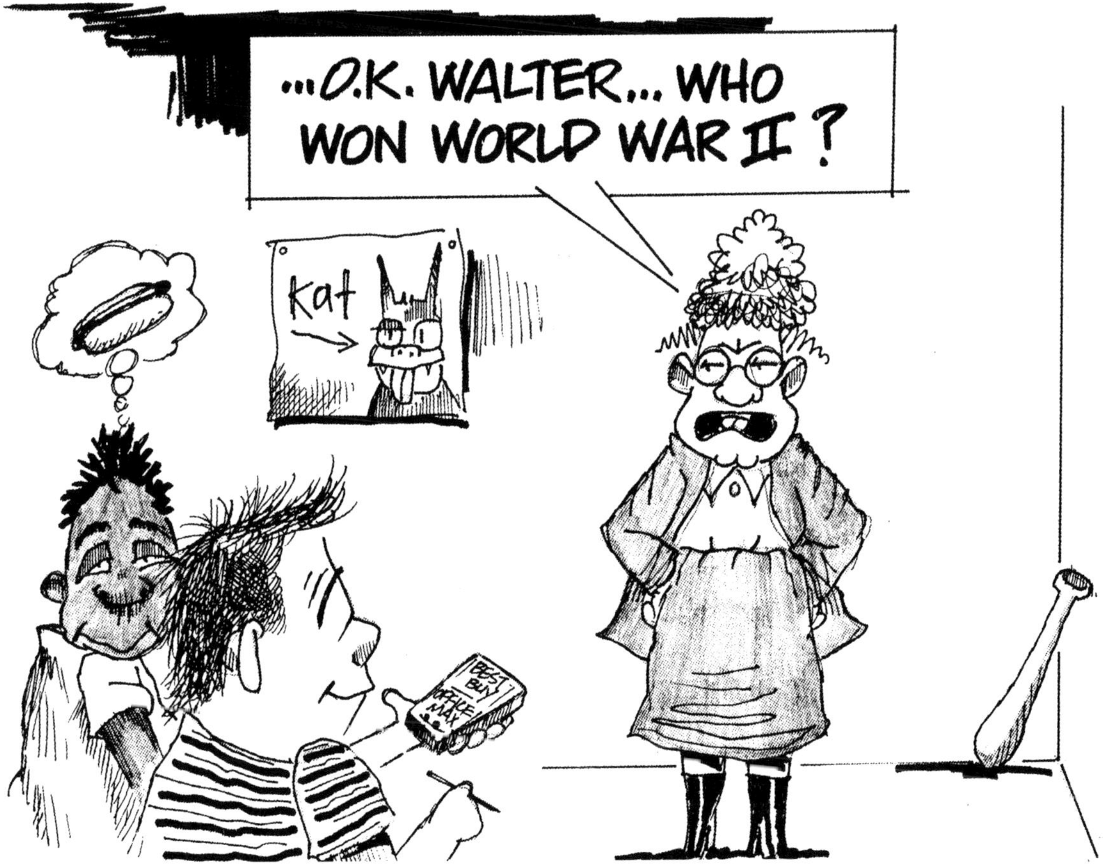
...O.K. WALTER... WHO WON WORLD WAR II ?
Kat
BEST BUY
OFFICE MAX

From the control room…

"If you get stung by a jellybean you have to pee on it." – Dodie

"It had two hide-a-couches and a tropicalgraphical map of the lake." – Purdy

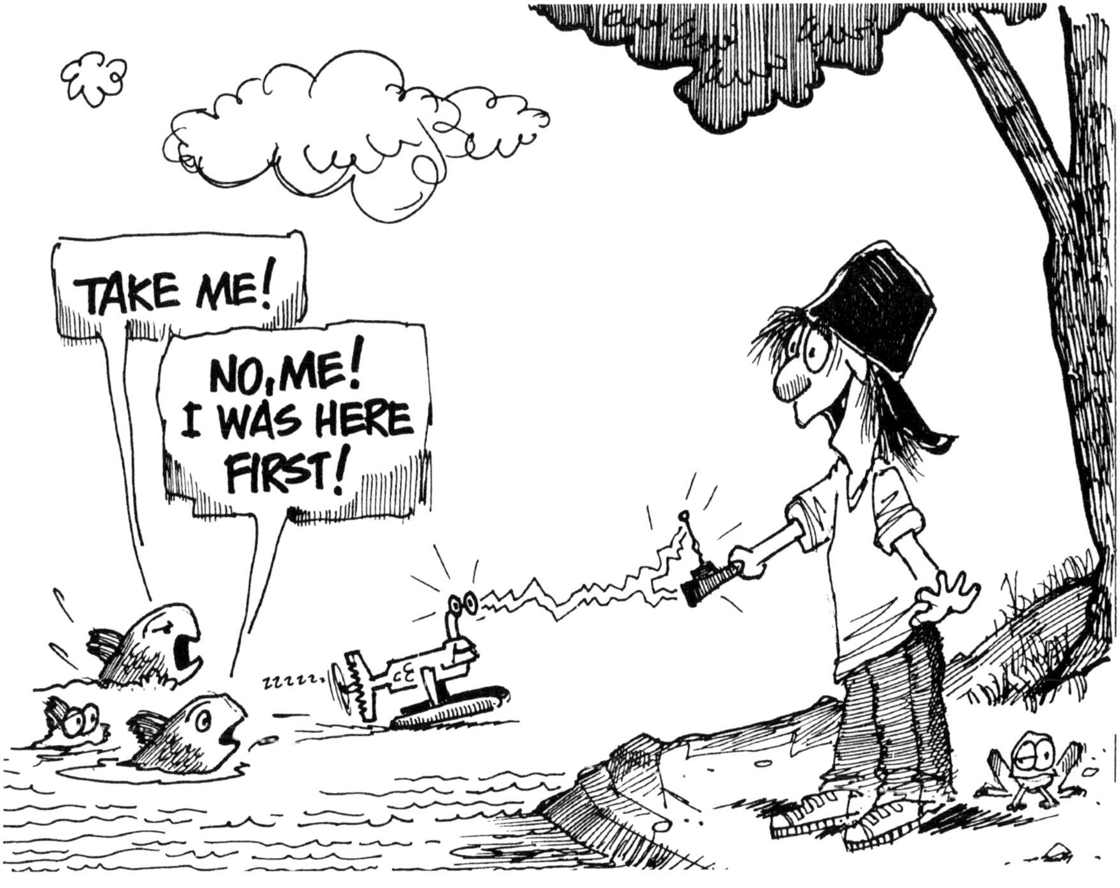
TAKE ME!
NO, ME! I WAS HERE FIRST!
ZZZZZ

From the control room…

"Kevin was squealing like a chicken." – Brad

"Now that I'm action area leaded, you guys can't blow me off!" – Ryan

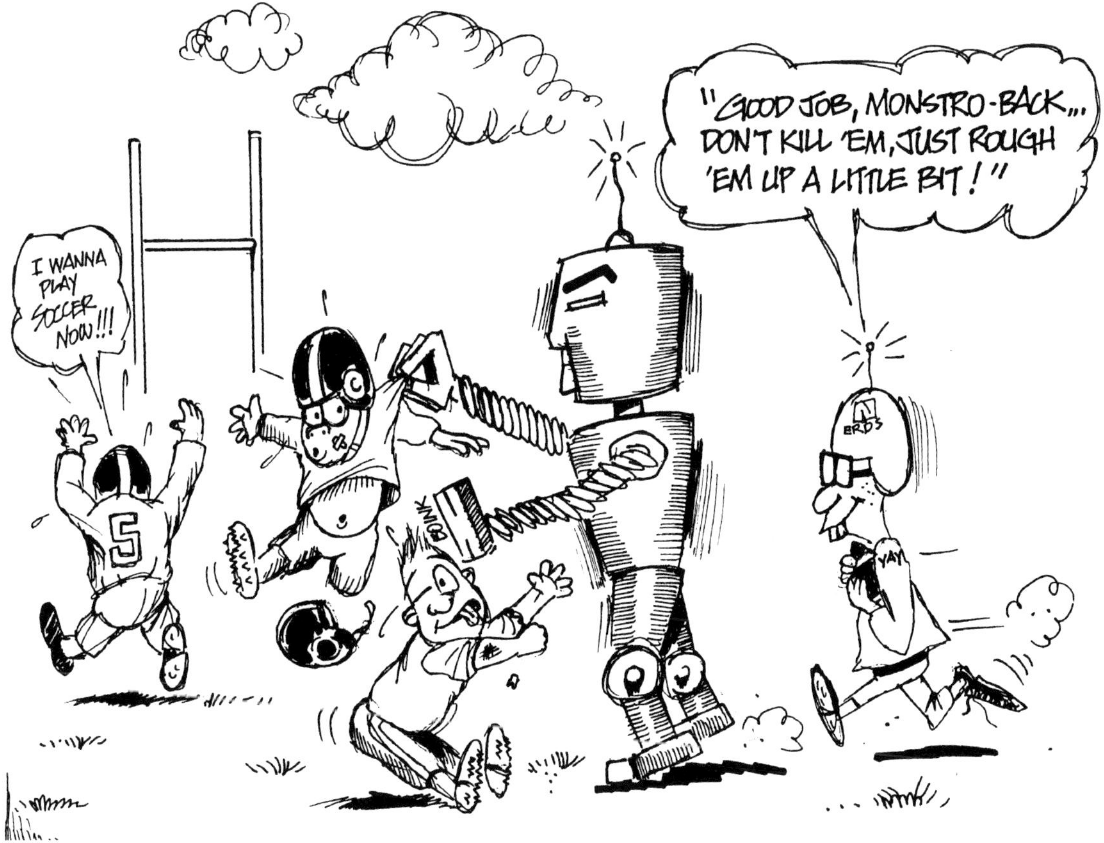

I WANNA PLAY SOCCER NOW!!!
"GOOD JOB, MONSTRO-BACK... DON'T KILL 'EM, JUST ROUGH 'EM UP A LITTLE BIT!"
BONK
YAY

From the control room...

"I don't get gas. My body absorbs it." – Sam

"On the weekends, there are 15 cars in the parking lot. You can count them on one hand." – Brian

"MY DAD'S GOTTA FIND THIS JOY STICK AND GPS LOCATOR FOR ICE CREAM WAY MORE FUN THAN HIS STEERING WHEEL AND SPEEDOMETER!"
ICE CREAM 12 MI.
NETSCOPE

From the control room...

"The turkey fur was flying!" – Kevin H.

"What size feet do you wear?" – Bodene

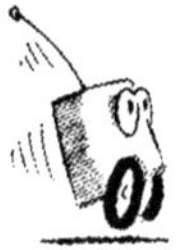

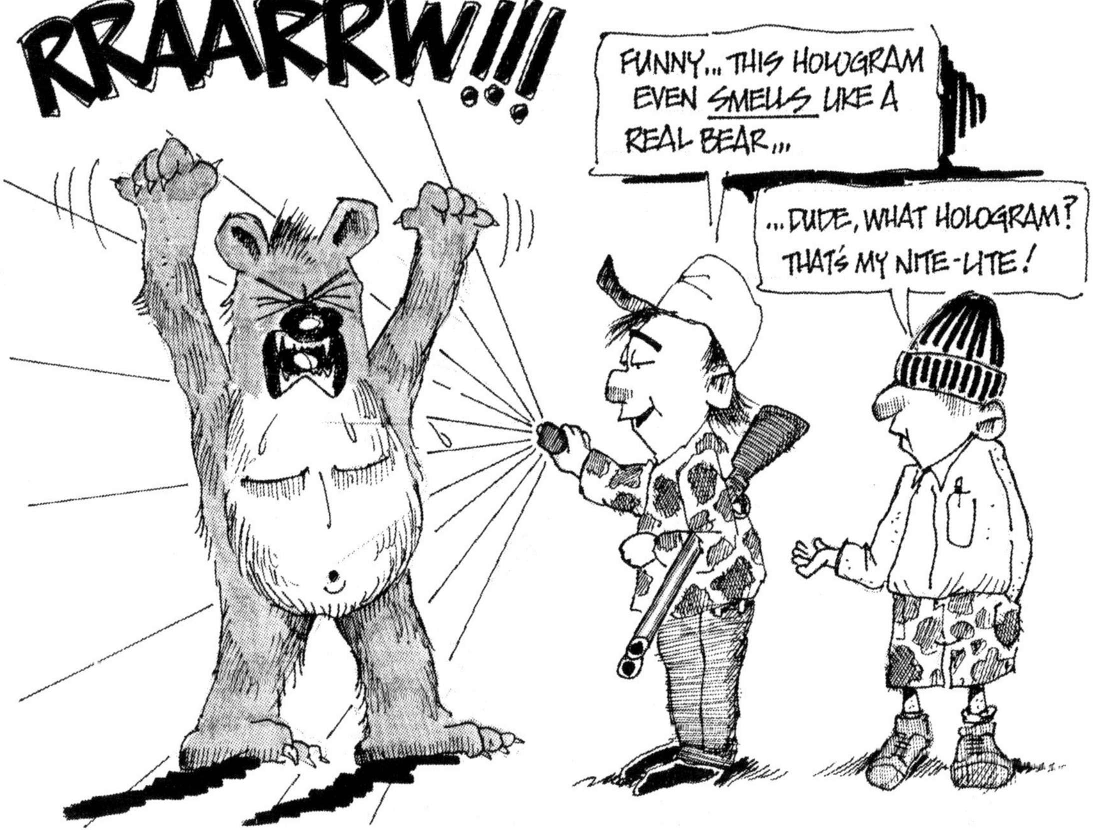
RRAARRRW!!!
FUNNY... THIS HOLOGRAM EVEN SMELLS LIKE A REAL BEAR...
...DUDE, WHAT HOLOGRAM? THAT'S MY NITE-LITE!

From the control room…

"He's not the sharpest bulb in the package." – Brian F.

"Do you guys want cheese on your cheeseburgers?" – Robbie

"I'M DOWNLOADING A GALLON!
...BRING the GIRLS... AND SOME DEODORANT!!!"
STATE
S
MOM
MAKE BOOZE YES NO
WHAT KIND BOURBON
SCOTCH
PROOF 80 86 90
QUANTITY PINT LITER

From the control room…

"From now on I'm going to have to be clericly correct." – Dirck

"I won't do that again twice!" – Dodie

"THANK YOU... DEAN... ON BEHALF... OF "STEWIE" MORGANKOWSKI... WHO... WAS UNAVOIDABLY DETAINED... I PROUDLY ACCEPT HIS... FREE GAME... I... MEAN... HIS... DIPLOMA!"
WU

From the control room...

"She's kinda from Missouri." – Sam

"You know, no matter what you say, men and women are different." – Dodie

"HUNK LOCATED... SATELLITE LOCKING ONTO LOCATION... PREPARE TO DOWNLOAD PREMARITAL DATA INFORMATION RECORDS... ...YOU GO, GIRL..."
Yours truly,
William Wallace
HUNK-LOCATER
DO NOT DISTURB

From the control room...

"I don't say stupid things, do I?" – Janet

"Either you are saying something about me or you are talking about me." – Purdy

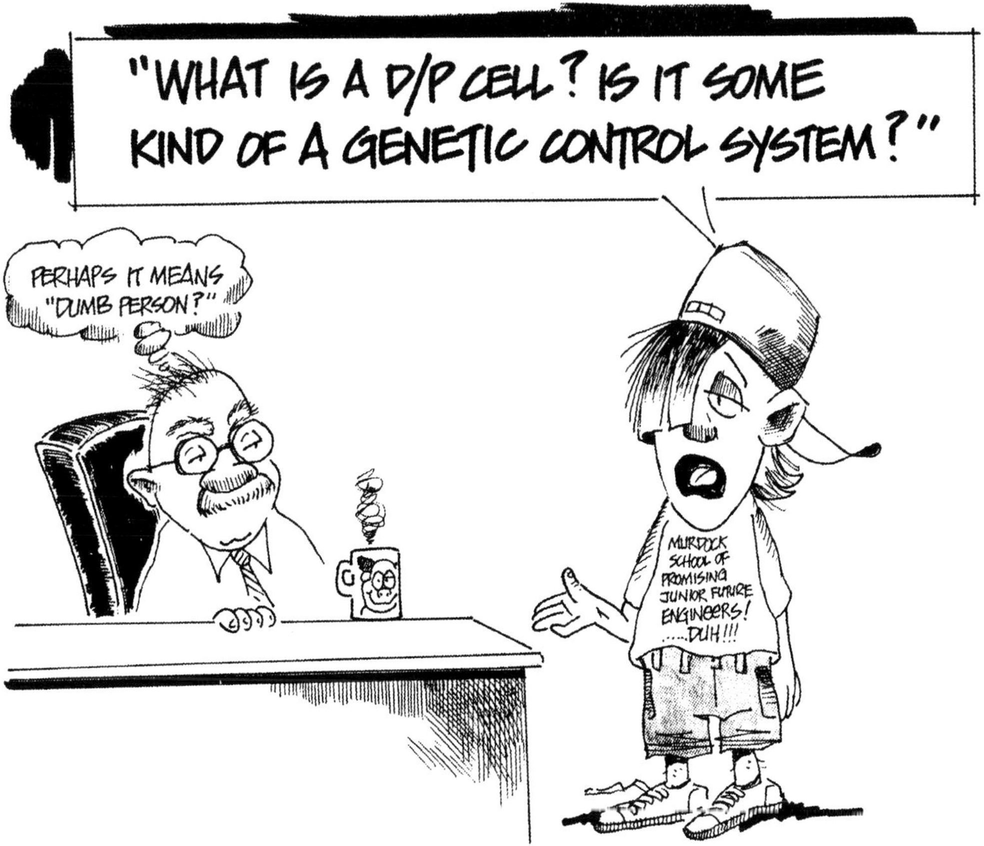
"WHAT IS A D/P CELL? IS IT SOME KIND OF A GENETIC CONTROL SYSTEM?"
PERHAPS IT MEANS "DUMB PERSON?"
MURDOCK SCHOOL OF PROMISING JUNIOR FUTURE ENGINEERS! ....DUH!!!

From the control room...

"We need to keep our minds to the grind stone." – Thad

"I've had the training. I am CRP certified." – Dub

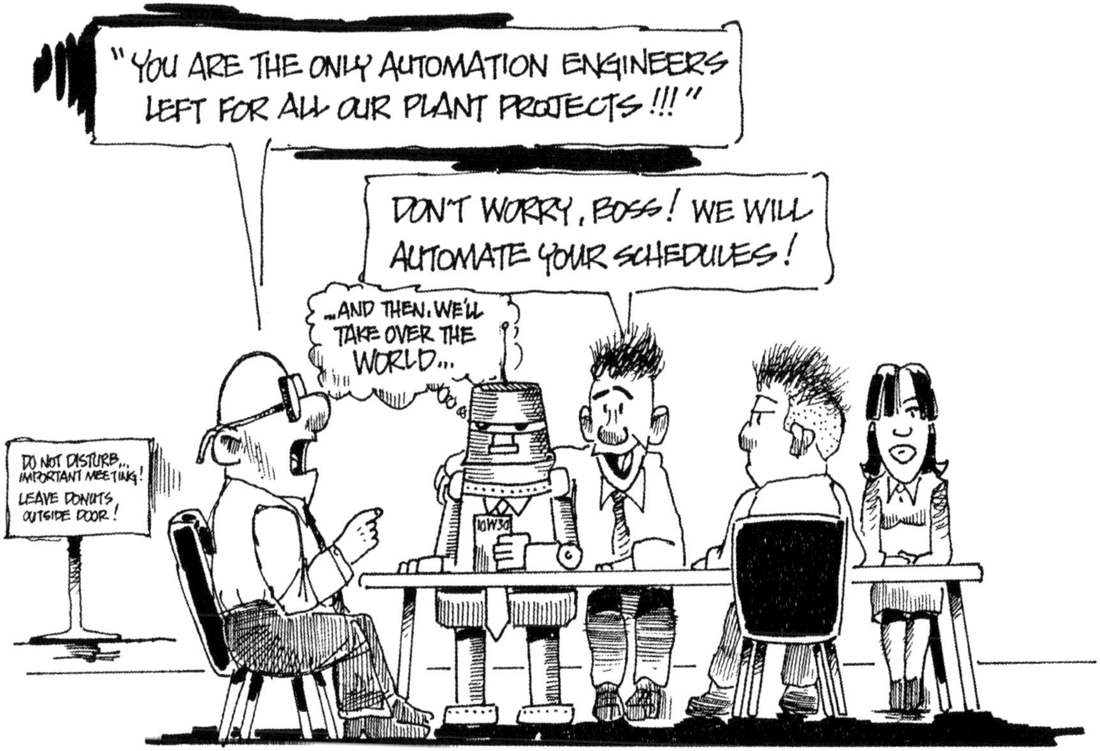
"YOU ARE THE ONLY AUTOMATION ENGINEERS LEFT FOR ALL OUR PLANT PROJECTS !!!"
DON'T WORRY, BOSS! WE WILL AUTOMATE YOUR SCHEDULES!
...AND THEN, WE'LL TAKE OVER THE WORLD...
DO NOT DISTURB... IMPORTANT MEETING! LEAVE DONUTS OUTSIDE DOOR!
10W30

From the control room...

Dub – "I've got to get my taxes done before my tax man leaves the country."
Thad – "Where is he going?"
Dub – "Colorado."

"...EXCUSE ME, FELLAS... WHERE ARE THE BODE PLOTS AND LAPLACE TRANSFORMS FOR EACH LOOP?"
"DO HUMANS... ALWAYS... BEHAVE... THIS... WAY?"
PLEASE! NO... NO MORE!
CONTROL THEORY
CONTROL THEORY VOL. 1

From the control room...

Don – "Did you just get here Matt?"
Matt – "Do you mean today or right now today?"

...THAT IS THE ACTUAL VALVE POSITION..... ...ISN'T IT ?
NO... IT IS THE CONTROLLER OUTPUT..., ISN'T IT ? WHY ARE THOSE HUMANS STILL DOING THAT?
VALVE POSITION HERE→ NO, HERE!
HERE!
CONTROL
...S-STOP! YOU'RE KILLING US!!!

From the control room...

"It's just money." – Thad

"A day here, a day there, that's 3 days vacation." – Uzi

HMMM... EXPONENTIALLY WASTING COMPANY TIME... PLAYING... VIDEO... GAMES... HUMANS ARE... SO CLEVER...
1504
3008
6016
64
188
376
752
2
4
8
16
32
SUPERKEYBOARD

From the control room...

"I got it wrong because I didn't circle the right answers." – Robbie

"There are some ripples in the waves." – Sonny

"MAYBE STRING HAS SOME DRAWBACKS AS THE UNIFIED FIELD THEORY..."
I HAVE TO BE MORE CAREFUL ABOUT WHO I "HELP" GET OUT OF GRAD SCHOOL...
MAGILICUDDY I ~AND SONS~ STRING INC.

From the control room...

"Two percent of 100 is a lot more than two percent of 15." – Dirck

"How much does 600 pounds of ammonia weigh?" – Scrap Iron

"THIS MUST BE AN EXAMPLE OF CONSERVATION OF SPIN!"
I HAD MORE FUN WHEN I WAS A VACUUM CLEANER...
NORTH AMERICAN MONITER
SOUTH AMERICAN MONITER
SILVER CORIOLIS is the BEST!
BLUE CORIOLIS is a lot BETTER!

From the control room...

John – "I think it's a lack of good genes."
Dodie – "I buy the expensive ones."

"...I DIDN'T THINK AN ANTI-MATTER LOGIC CONTROLLER WAS A GOOD WAY OF ADDING DIVERSITY TO A SAFETY INTERLOCK SYSTEM..."
PLANT NO. 1
KABLAM!
UH..., I WAS OFF THAT DAY!
CAN I HAVE YOUR OFFICE?

From the control room...

"I've got a different way of versing the American language." – Dirck

"I got in trouble for not dotting my tee's and crossing my I's." – Lucas

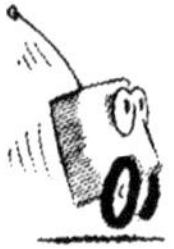

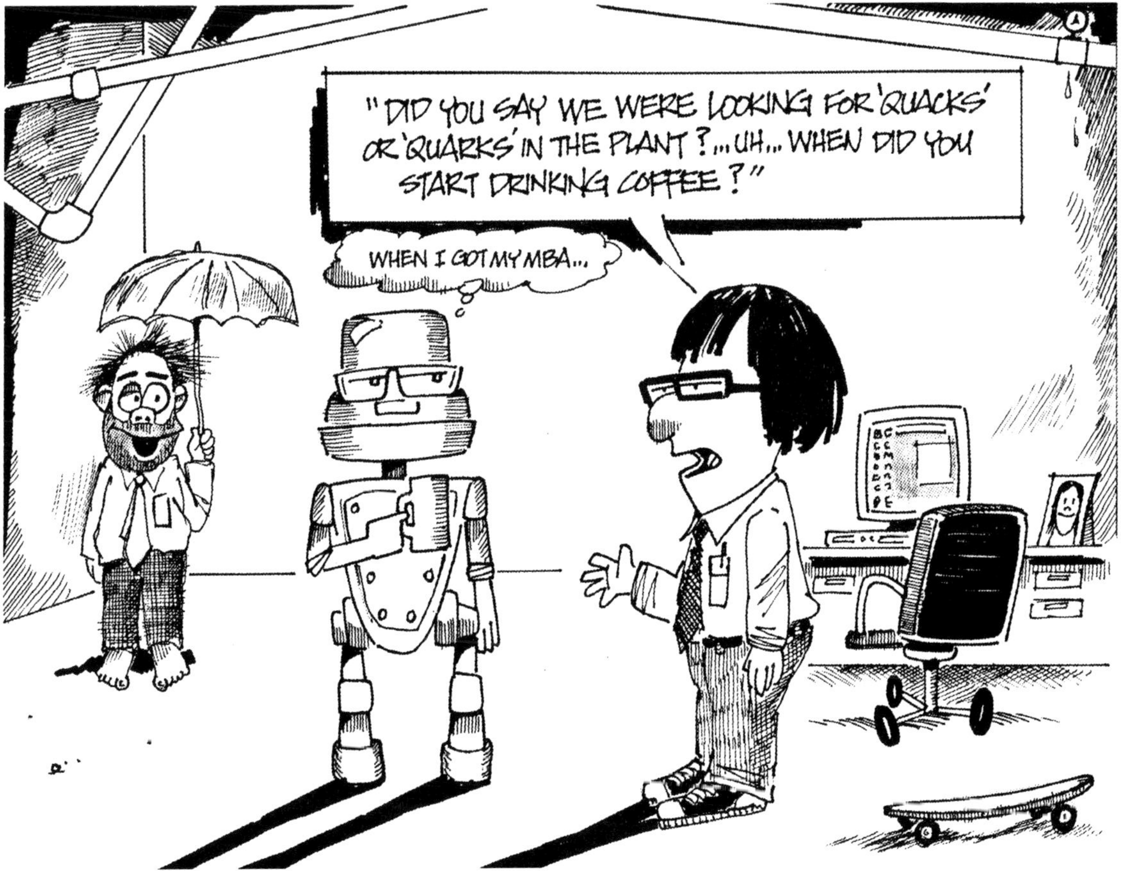
"DID YOU SAY WE WERE LOOKING FOR 'QUACKS' OR 'QUARKS' IN THE PLANT ?... UH... WHEN DID YOU START DRINKING COFFEE ?"
WHEN I GOT MY MBA...

From the control room…

"I was just thinking to myself, out loud." – Ward

"These are only going to be one answer words." – Dirck

QUIET! CERTIFICATION OF AUTOMATION PROFESSIONALS EXAM TODAY! QUIET!
pssst... ...Remember... I built you... what's a D/P cell?
FAMOUS ISA
People!
ISA ON
ISA!
HOW TO BUM MODEL AIRM
BIG BOOK OF ISA STUFF
ISA RECIPES
ISA
RUPERT MURDOCK
ISA IN THE OLD DAYS
COWS
WHERE IS ISA?
ISA TRIVIA
CD
I KNOW
I DON'T KNOW

From the control room...

"This is not user easy." – Sam

"Give it a second. It might take a few minutes." – Dodie

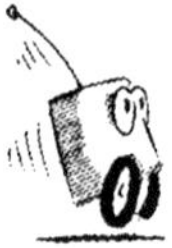

"YOU ARE DOING A GREAT JOB, BUT UNFORTUNATELY YOU ARE CONTROLLING THE VIRTUAL PLANT INSTEAD OF THE REAL PLANT!"
WOULD YOU LIKE ME TO ADJUST THE TEMPERATURE IN HERE? I CAN DO IT WITHOUT GETTING UP, YOU KNOW...
ON OFF
VOLUME
CD-
DVD-

From the control room…

"Getting into that Escort Wagon was like getting into a Vette." – Cole

"I don't want a Corvette. I want a Lambordini." – Dub

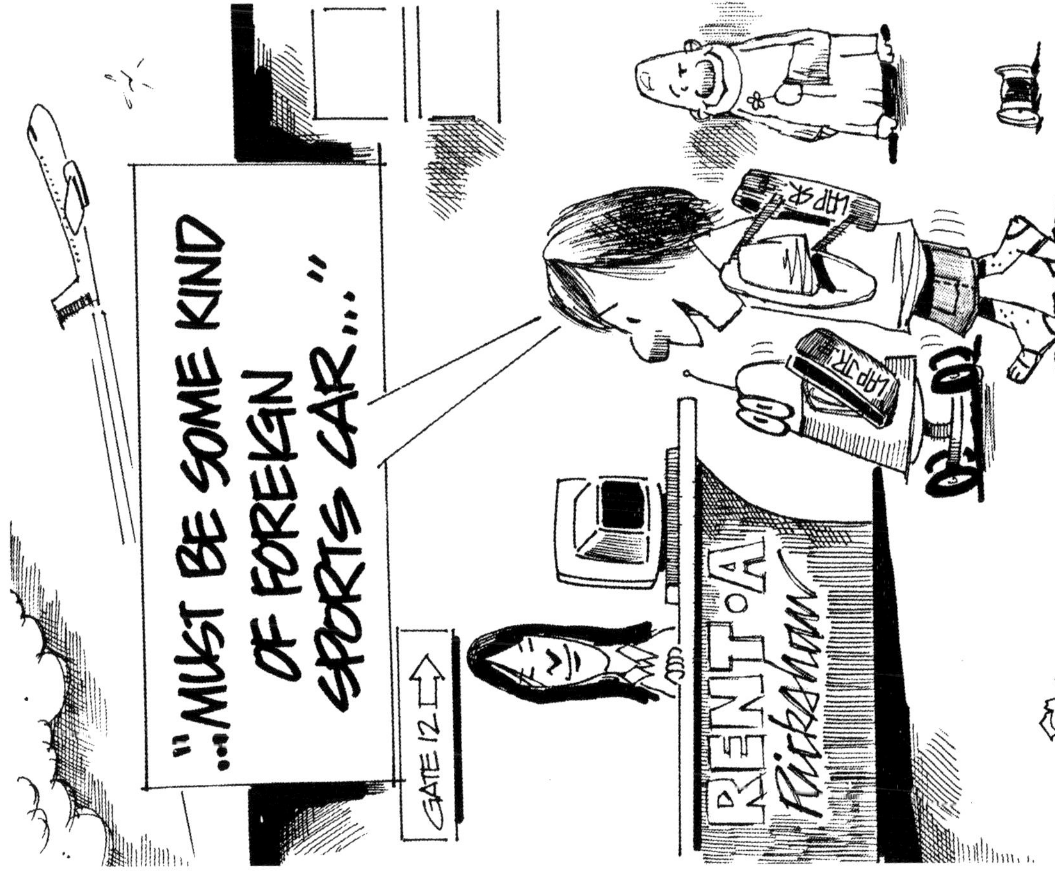
"...MUST BE SOME KIND OF FOREIGN SPORTS CAR..."
GATE 12
RENT·A Rickshaw
LAP SR.
LAP JR.

From the control room…

"What time is our 2 o'clock meeting?" – Mike S.

"The agitator is wired backwards so we are unstirring instead of stirring." – Dub

"HE'S ASKING IF THE RED BUTTON STARTS IT OR STOPS IT..."
...OH MY... I KNOW WHERE THE COFFEE POT IS!
PLANT start up TODAY!

From the control room...

"They're going to pay 3 digit money, like $100,000 a year." – Janet

"You'll get an incentive for eight twelfths or three quarters of the year." – Steve A.

Country A
Country B
CONFIGURATION ENGINEERS... 4.99 HR.
"...NOW WHAT?"
CONFIGURATION ENGINEERS 9.99 HR.
...HOW WOULD I KNOW? YOU'RE THE ONE WITH THE MBA...

From the control room…

Quinton – "I was outside helping Dave when a bird crapped on me."
Bodene – "Dave Crull?"
Quinton – "No, the bird."

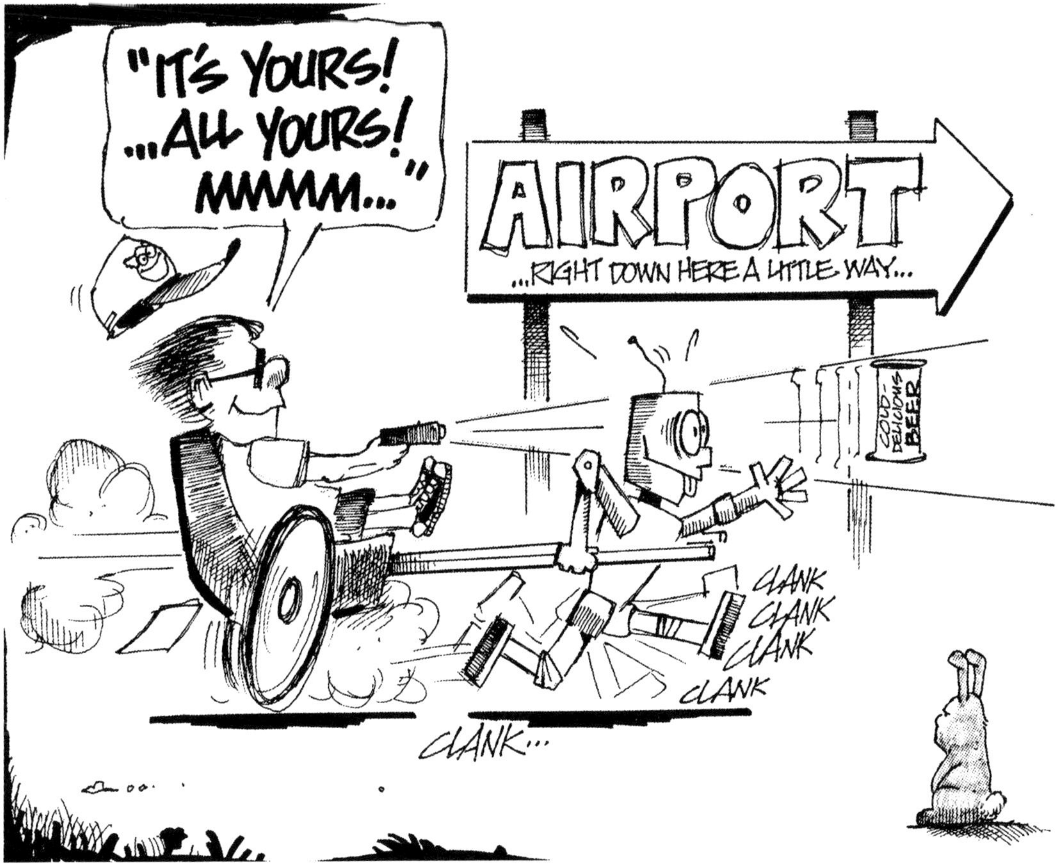
"IT'S YOURS!
...ALL YOURS!
MMMM..."
AIRPORT
...RIGHT DOWN HERE A LITTLE WAY...
COLD-DELICIOUS BEER
CLANK
CLANK
CLANK
CLANK
CLANK...

From the control room…

"It's like when you are trying to find your Minerva." – Cole, meaning Nirvana

"Is Apocalypse the three headed lady?" – Quinton

IT MUST HAVE BEEN SOMETHING I DID IN A PAST LIFE OR SOMETHING!
BOSS
DEAR YOUNG ENGINEER: YOU ARE SIGNED UP FOR HINDU 1.2. AND 3 DON'T TRY TO GET OUT OF IT. HAVE A WONDERFUL DAY!
THE BOSS
HINDU... THE EASY WAY!

From the control room...

"The squeaky gets the worm." – Keith

"He looked like the bird that swallowed the canary." – Jay

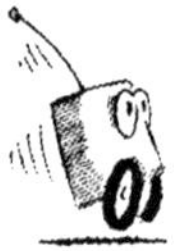

...WILL I BE NEEDING AN ELEPHANT? WHO IS SANJI? WHY MUST I LEARN HINDU?
HEY EINSTEIN! GOT ANY WATER?
SANJI'S ELEPHANT RENTALS! DISCOUNTS TO YOUNG ENGINEERS!

From the control room…

Don – "You're 5' 11" aren't you?"
Robbie – "No, I'm 5' 12"."

...DOESN'T HAVE MUCH OF A SENSE OF HUMOR, DOES HE?
DEAR YOUNG ENGINEER... FRATERNIZATION WITH MONKEYS ON STARTUPS IS HIGHLY DISCOURAGED... the BOSS
DEW
ATLAS OF MAPS

From the control room…

"Somebody's got to be the center of attraction." – Radar

"They will never goat me into doing that." – Purdy

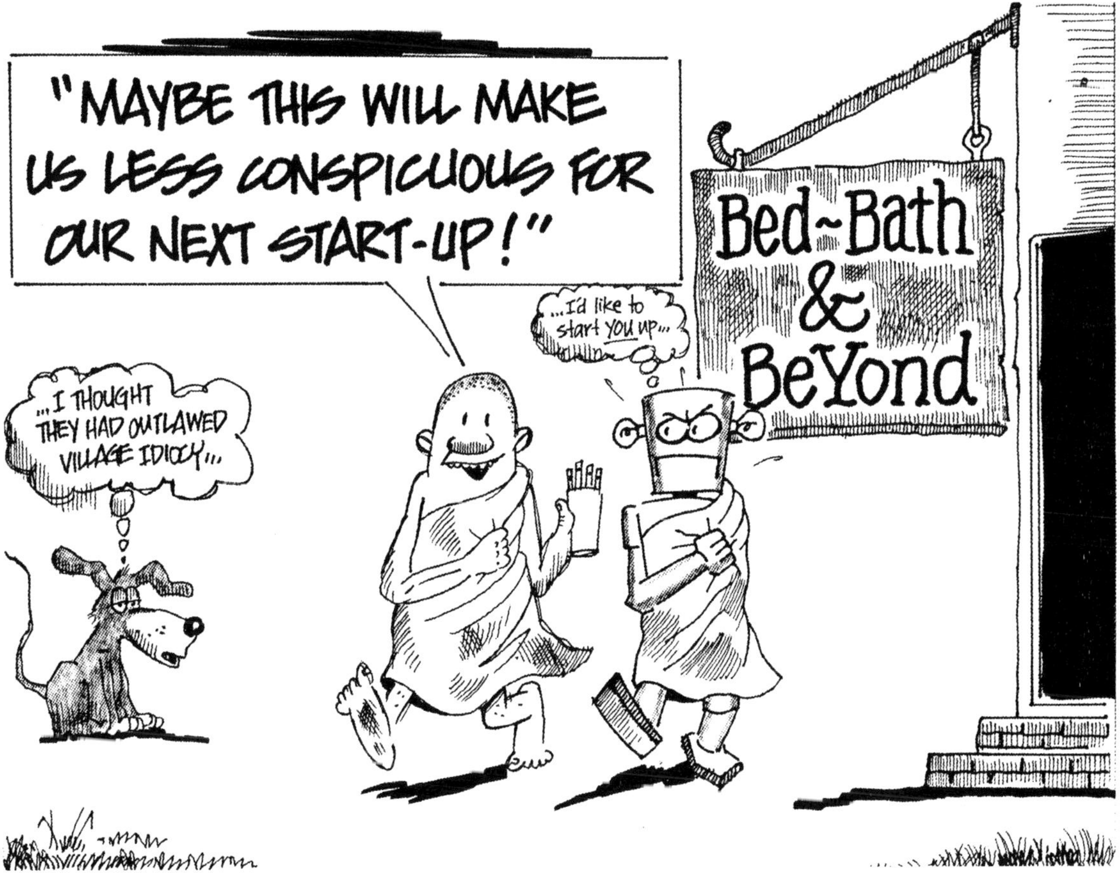
"MAYBE THIS WILL MAKE US LESS CONSPICUOUS FOR OUR NEXT START-UP!"
...I'd like to start YOU up...
Bed~Bath & BeYond
...I THOUGHT THEY HAD OUTLAWED VILLAGE IDIOCY...

From the control room…

"They keep 80 percent of the product and give the other 30 percent away." – Kevin H.

"This place only runs 6 months out of the month." – Thad

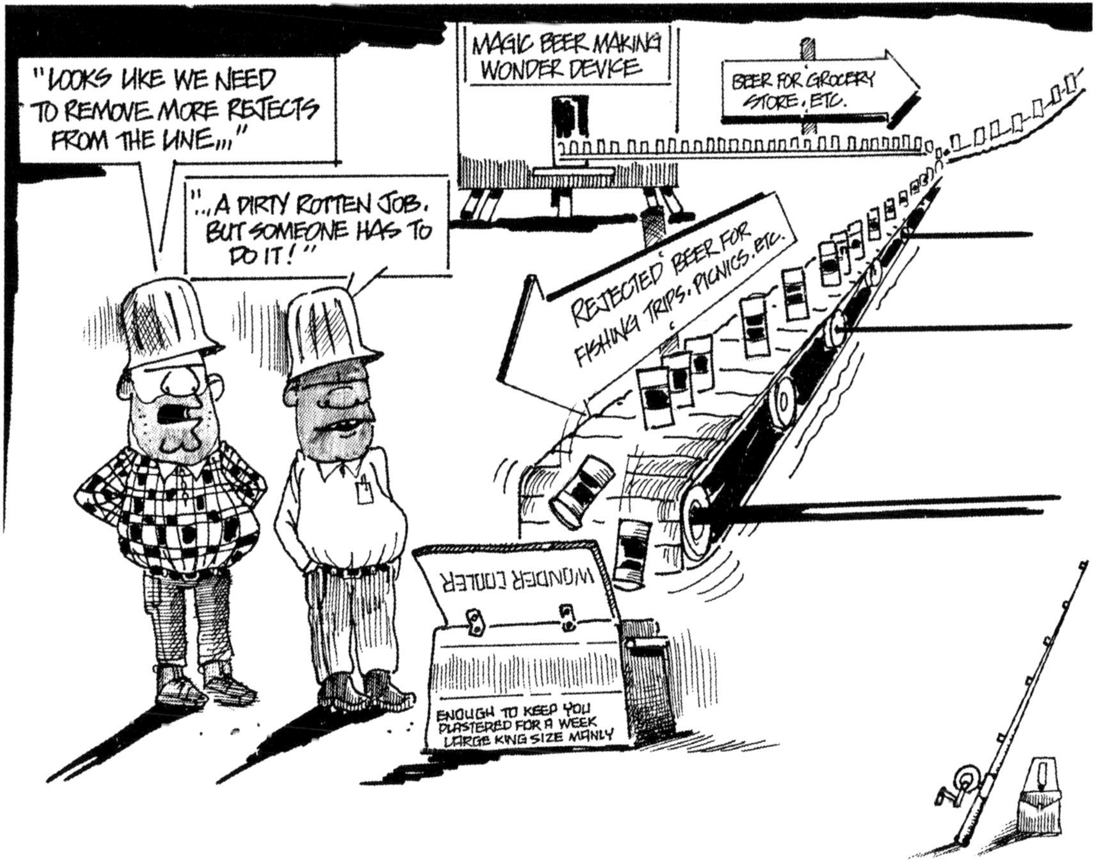
"LOOKS LIKE WE NEED TO REMOVE MORE REJECTS FROM THE LINE,,,"
"...A DIRTY ROTTEN JOB. BUT SOMEONE HAS TO DO IT!"
MAGIC BEER MAKING WONDER DEVICE
BEER FOR GROCERY STORE, ETC.
REJECTED BEER FOR FISHING TRIPS, PICNICS, ETC.
WONDER COOLER
ENOUGH TO KEEP YOU PLASTERED FOR A WEEK
LARGE KING SIZE MANLY

From the control room…

Kevin H. – "What's Jerry's last name?"
Quinton – "Jerry who?"

"IS THAT MR. HOWELL?!!"
...THERE GOES OUR CUSTOMER SATISFACTION RATING...
HOWELL ROBOTICS
"Where ROBOTS RULE"

From the control room…

"I'm fair but I'm rigid." – Dana

"I was in the throngs of passion." – Dirck

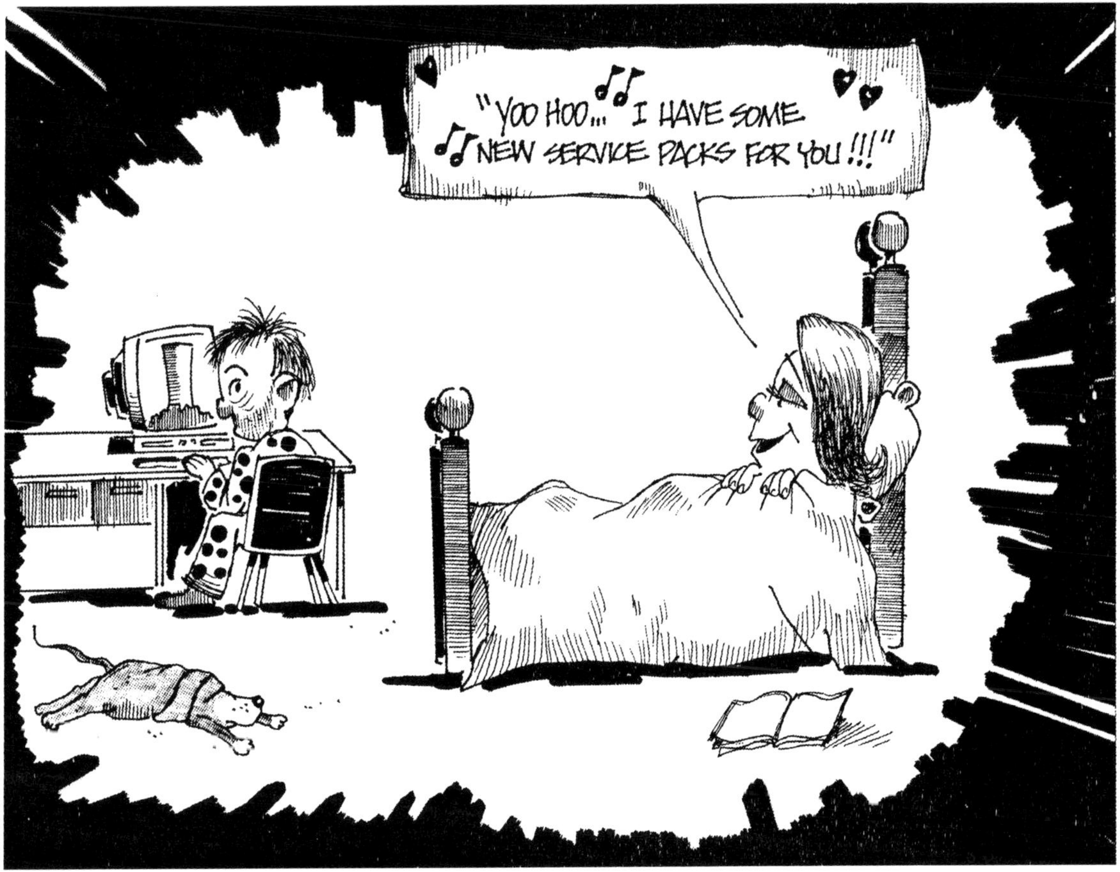
"YOO HOO... I HAVE SOME NEW SERVICE PACKS FOR YOU!!!"

From the control room...

"This is one of those airless all electric valves." – Purdy

"It doesn't smell like transmission fluid and it has a different velocity." – Robbie

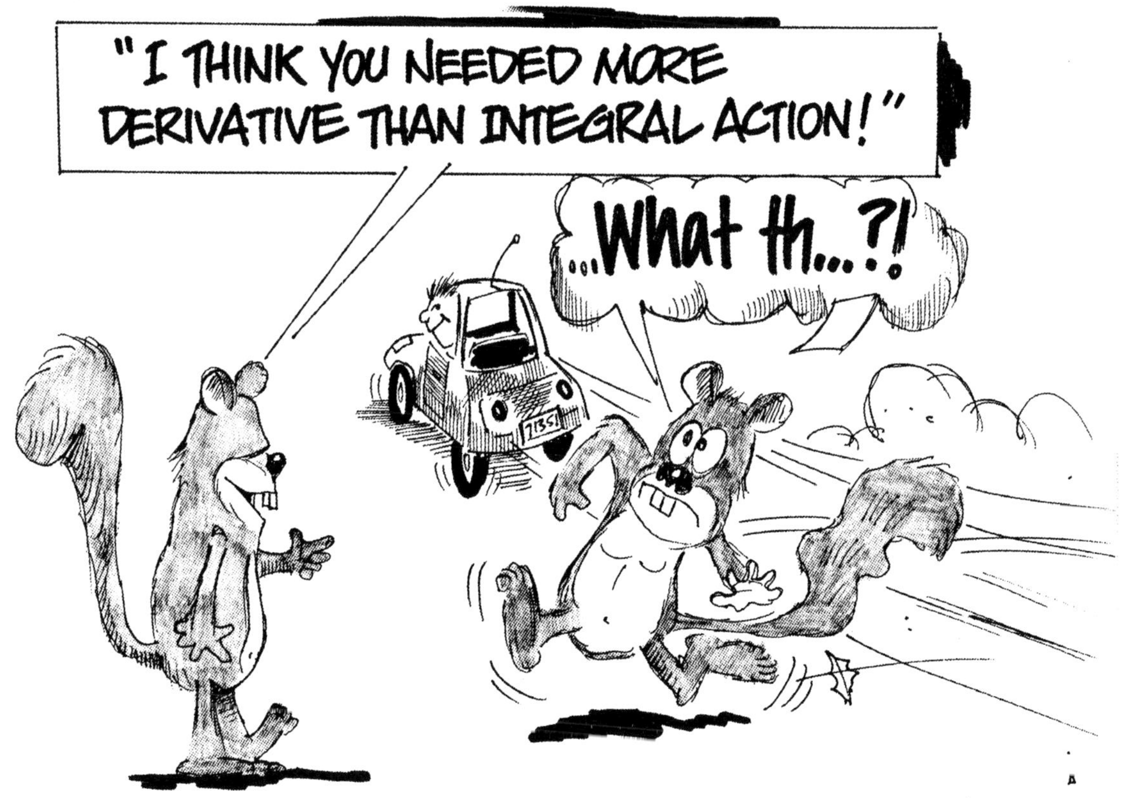
"I THINK YOU NEEDED MORE DERIVATIVE THAN INTEGRAL ACTION!"
...What th...?!

From the control room...

"This is humidity rain." – Dana

"The pump has 23 umps." – Purdy

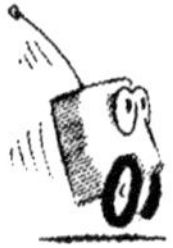

WHAT WAS THE GREATEST INNOVATION FOR REDUCING CAPITAL PROJECTS IN THE '60's?
"...MY GUESS IS THE INTRODUCTION OF MOUNTAIN DEW!"
BREAK ROOM
LOT O' DEW

From the control room...

"I'm saying the same thing but it's different." – Brillo

"ROE means Return On Investment." – Purdy

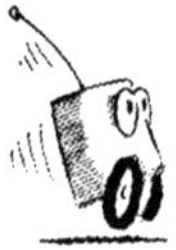

"...LIKE, THE DUDE IS STICKING IT CLEAN AND TOTALLY DIALED IN ON SURFING THE NET WITH A PHAT COMPUTER IN A TOTALLY RAD CUBE!"
...I GUESS I'M OLDER THAN I THOUGHT...
MAGNUM V.S. the "MOSQUITO" BOINK!
W.U. School of PROMISING HIP YOUNG ENGINEERS!
DEW

From the control room...

"I weigh 205 cold naked." – Bodene

"Is that your frozen ice?" – Thad

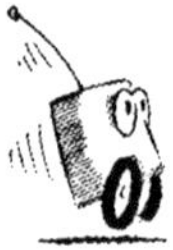

"I-I-I UN-UN-UNDERST-STAND TH-TH-THE HVAC E-ENGINEER IS F-F-F-FROM ALASKA!"
...AND QUITE MAD, THEY SAY.
COMPLITE

From the control room…

"You stay here and go home." – Bodene

"My mom is related to her but I'm not." – Sam

"APPARENTLY, EXTRATERRESTRIALS AREN'T SCREENED OUT BY OUR NEW SECURITY SYSTEM!"
KF NEWN ALL KREIN BEN? (I AM A MBA... WHERE IS THE FRONT OFFICE?)
BF
MEOW!

From the control room…

"You can tell a lot about a bird by looking at his butt." – Brad

"Are you a hydroconrick?" – Dirck

"...I'M NOT A GEEK!"
"AND I'M NOT A ROBOT!"
FOR ULTIMATE SECURITY, ALL GEEKS AND ROBOTS ARE DENIED ACCESS! HAVE A NICE DAY...
TOP SECRET
JUST REGULAR SECRET
ANYBODY CAN LOOK IN HERE...

From the control room…

"There's a flicker at the end of the tunnel." – Howes

"His pay will be retrofitted." – Robbie, talking about back pay

THE UNION SAYS THAT YOU CAN'T REPLACE 10 OLD ENGINEERS WITH A LAP TOP, EVEN IF IT DOES HAVE A 21 INCH SCREEN...
Well... can I get rid of YOU?!
!

From the control room...

"I just got my thing cut in half." – Purdy

"You had that funny look on your eyes." – Janet

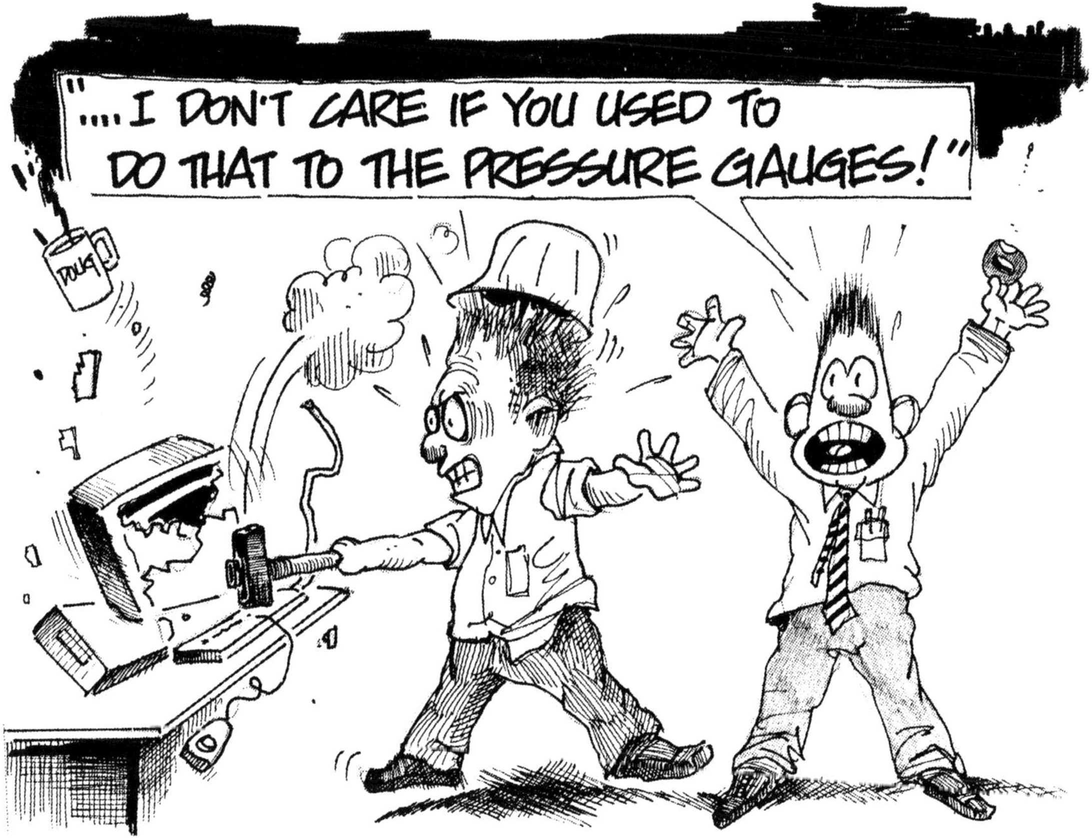
"....I DON'T CARE IF YOU USED TO DO THAT TO THE PRESSURE GAUGES!"
DOUG

From the control room...

"That's one of those rubber booty things." – Uzi, explaining an expansion joint to Dodie

"If you want it, just go and buy it. That's my motto." – Thad

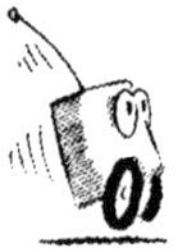

THAT'S WHY WE BUY THE "EVERY ANGLE" TYPE...
"DO YOU HEAR SOMETHING? A SQUEAK, MAYBE?

From the control room…

"Have you ever had hair?" – Jenny, talking to Dirck

"Rod is infringing on my pattern rights." – Fred

...I DON'T CARE IF EVERYONE HAS HEARING PROTECTION... STOP PLAYING RAP AND HEAVY METAL IN THE CONTROL ROOM!
OH MAN...

From the control room…

"I could start making snard comments." – Dodie

"We all know who runs the pants in your family." – Purdy

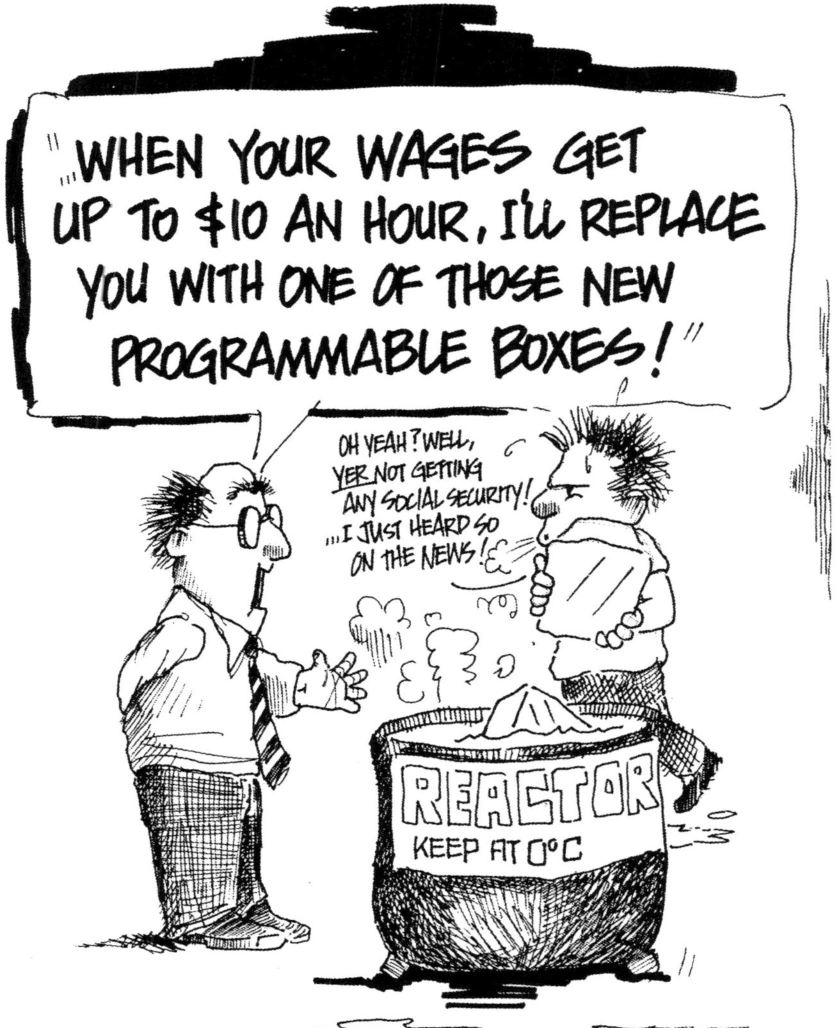
"...WHEN YOUR WAGES GET UP TO $10 AN HOUR, I'LL REPLACE YOU WITH ONE OF THOSE NEW PROGRAMMABLE BOXES!"
OH YEAH? WELL, YER NOT GETTING ANY SOCIAL SECURITY! ...I JUST HEARD SO ON THE NEWS!
REACTOR
KEEP AT 0°C

From the control room...

"Fred, how do you spell caulk?" – Sam

"Meeker has three disks that are inflamminized." – Purdy

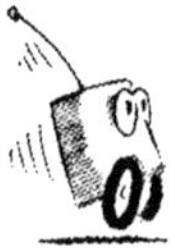

IF SPELLING ISN'T IMPORTANT, CAN I WRITE THE INSTRUCTION BOOK IN EXCEL?
MMDLGRK...

From the control room...

"Are those steel toed shoes?" – Jeff

"Define steel toed shoes." – John Powell

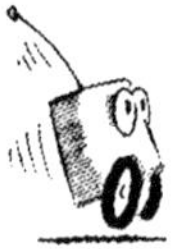

HEY BOSS! LOOK WHAT I DID! I WROTE A PROGRAM THAT DOWN-LOADS MY CREDIT CARD ACCOUNTS DIRECTLY INTO MY EXPENSE STATEMENTS! ...BOSS? BOSS?
?

From the control room…

"Poor Eric. He works himself to the fingers." – Janet

"If you have time to lean you have time to clean." – Thad

I TOLD YOU BEFORE TO STOP PUTTING MOVING SMILEY FACES ON ALL THE SCREENS!
PREZ

From the control room...

"Thad, your hat looks like one from Irabia!" – Dirck

"Anything you buy you have to purchase." – Purdy

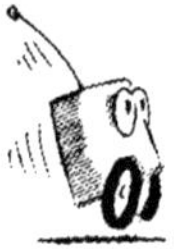

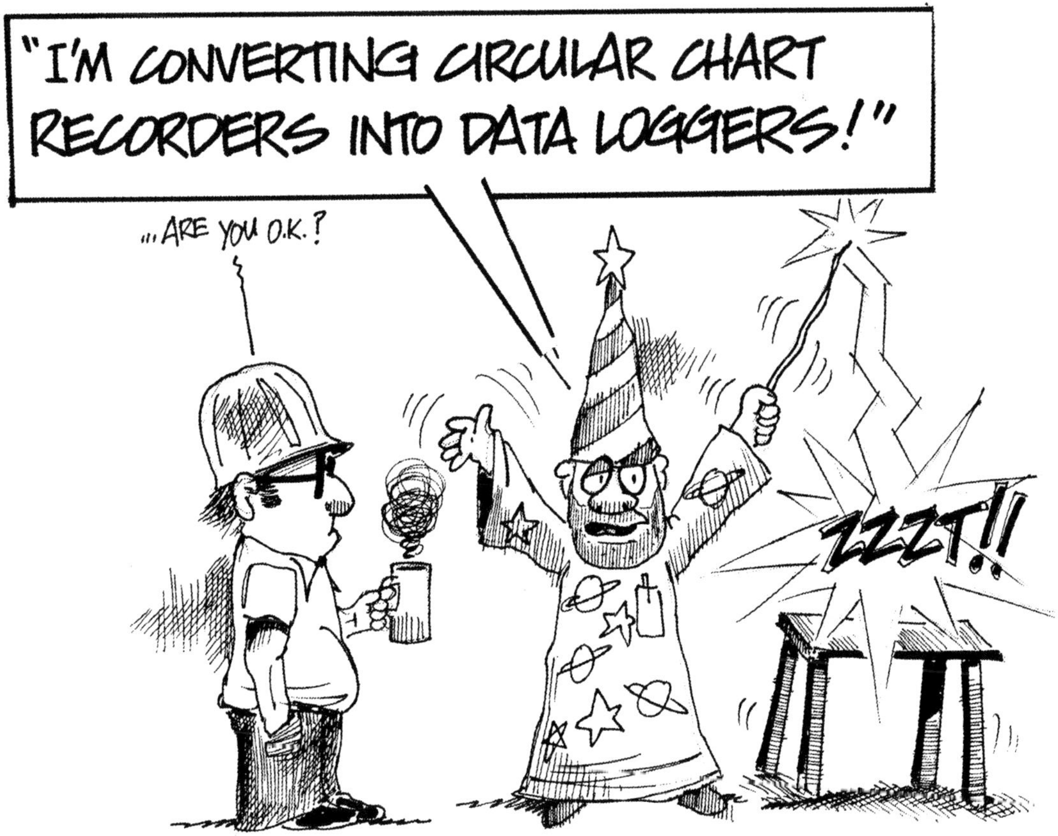
"I'M CONVERTING CIRCULAR CHART RECORDERS INTO DATA LOGGERS!"
...ARE YOU O.K.?
ZZZT!!

From the control room...

"How come we took that valve off the handle?" – Cole

"Is it supposed to be doing that?" – Diane, as water dripped down from the control room ceiling

UMMM... AUTOMATION DOES NOT MEAN PUTTING LARGER HANDLES ON THE VALVES...
...SAYS WHO?
LARGE HANDLE

From the control room...

"I made cookies but they are not mine." – Dodie

"I even took piano lessons for seven years." – Dirck

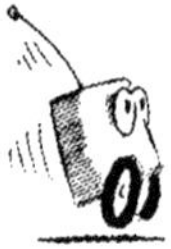

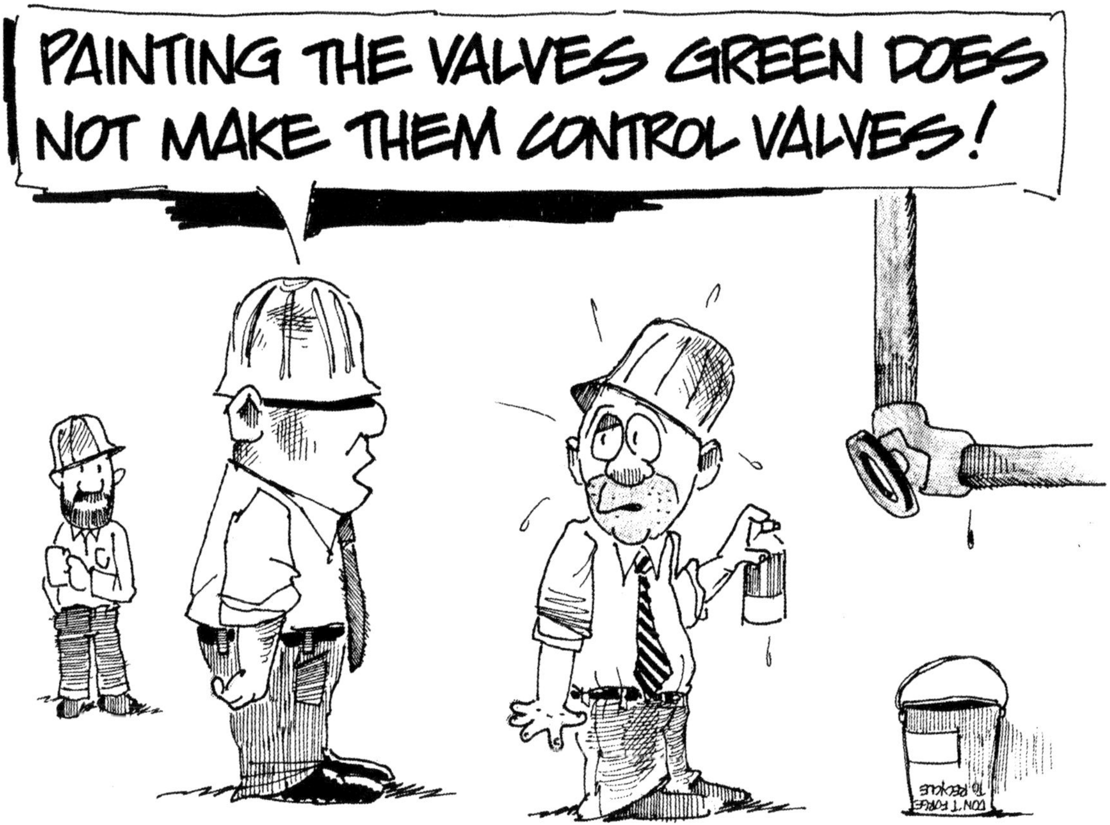
PAINTING THE VALVES GREEN DOES NOT MAKE THEM CONTROL VALVES!

From the control room...

"She was working with a temp of crews." – Brian W.

"This condensate is like boiling water." – Robbie

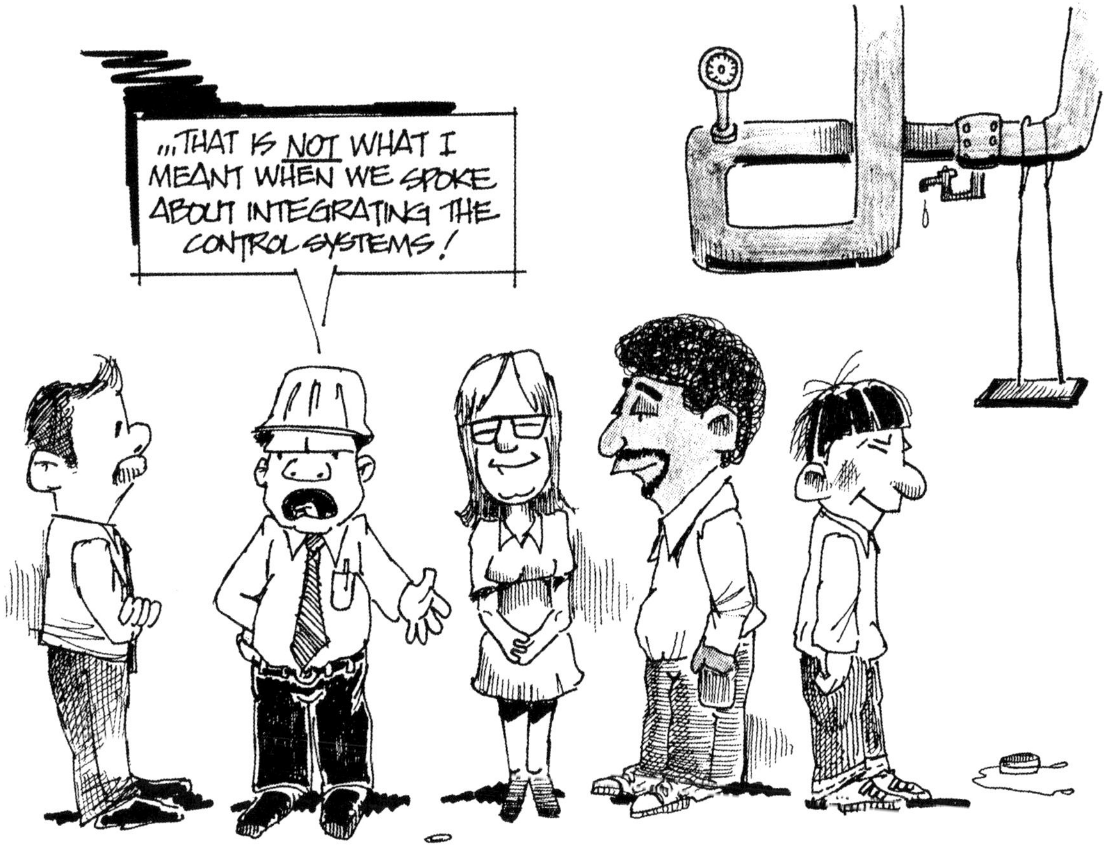
...THAT IS NOT WHAT I MEANT WHEN WE SPOKE ABOUT INTEGRATING THE CONTROL SYSTEMS!

From the control room…

"I wonder if my ex-nephew in law still works there." – Dodie

"Jim, I need you to order your time card." – Cole

"...LET ME EXPLAIN THE ORGANIZATION CHART TO YOU. THERE ARE 156 ENGINEERS... ...AND ME."
LUNCH..
ME!
YOU

From the control room...

"We've got over 100 hours of pog time in 3 days." – Cole

"The pipe supports filled up with ice and cracked." – Sonny

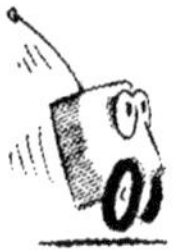

"IT'S NOT A BAD PLACE TO WORK...
EXCEPT FOR THE DIRT..."
"O.K..O.K., WE'LL PUT A NEW
FILTER IN THE A/C UNIT..."

From the control room…

"Everyone can breath easier, the butt can is cleaned out." – Dodie

"Yeah, Bodene is probably sitting in a naked chair." – Thad

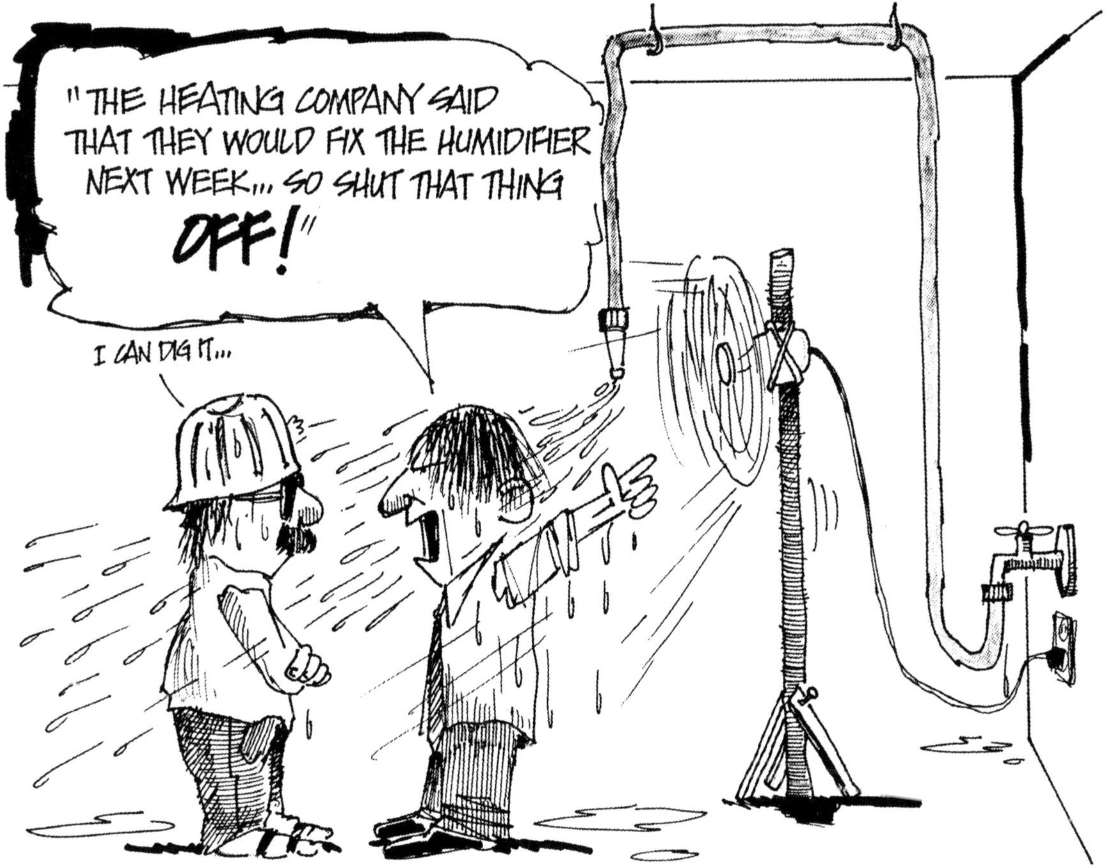
"THE HEATING COMPANY SAID THAT THEY WOULD FIX THE HUMIDIFIER NEXT WEEK... SO SHUT THAT THING OFF!"
I CAN DIG IT...

From the control room...

"I thought you had to use them before they were dead." – Dodie, talking about organ transplants

"Remember the bi-centennial in 1969?" – Cole

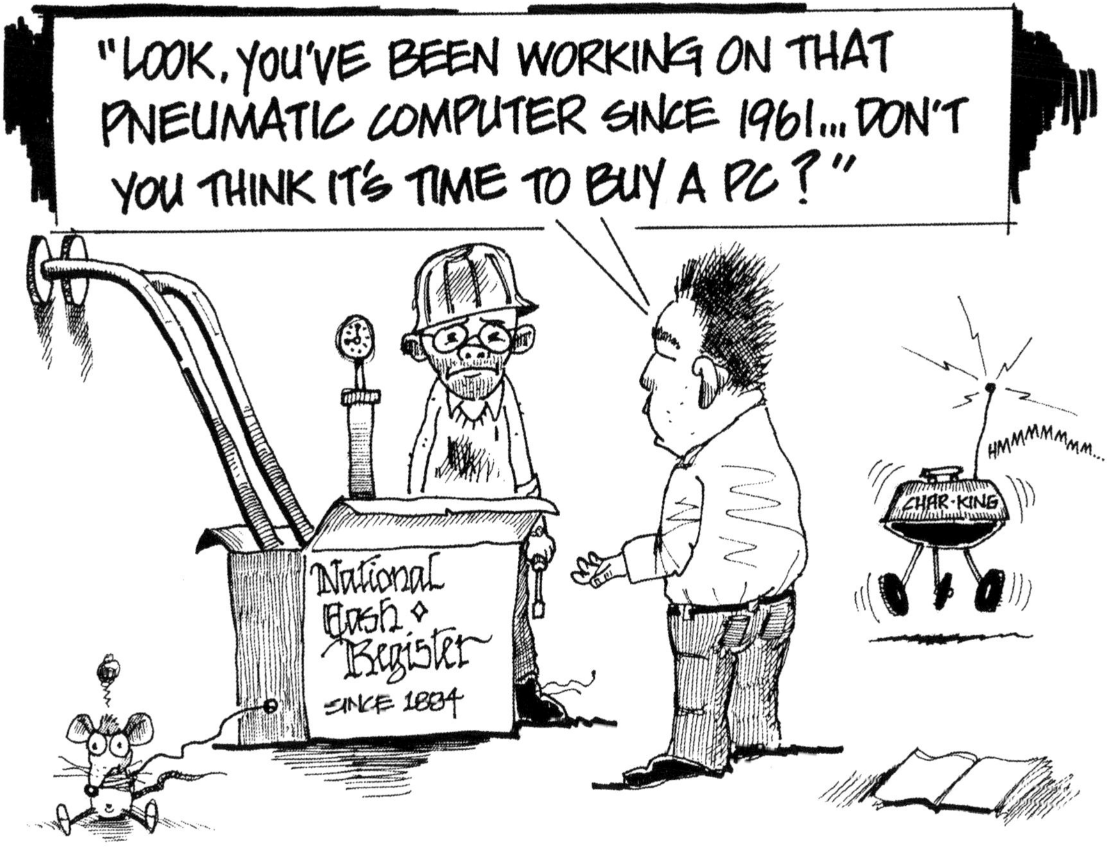
"LOOK, YOU'VE BEEN WORKING ON THAT PNEUMATIC COMPUTER SINCE 1961... DON'T YOU THINK IT'S TIME TO BUY A PC?"
National Cash Register
SINCE 1884
CHAR-KING
HMMMMMMMM...

From the control room…

"Why didn't you tell us this was Ugly Day Hat?" – Purdy

"Thirty dollars of my pocket is coming out of my money." – Lucas

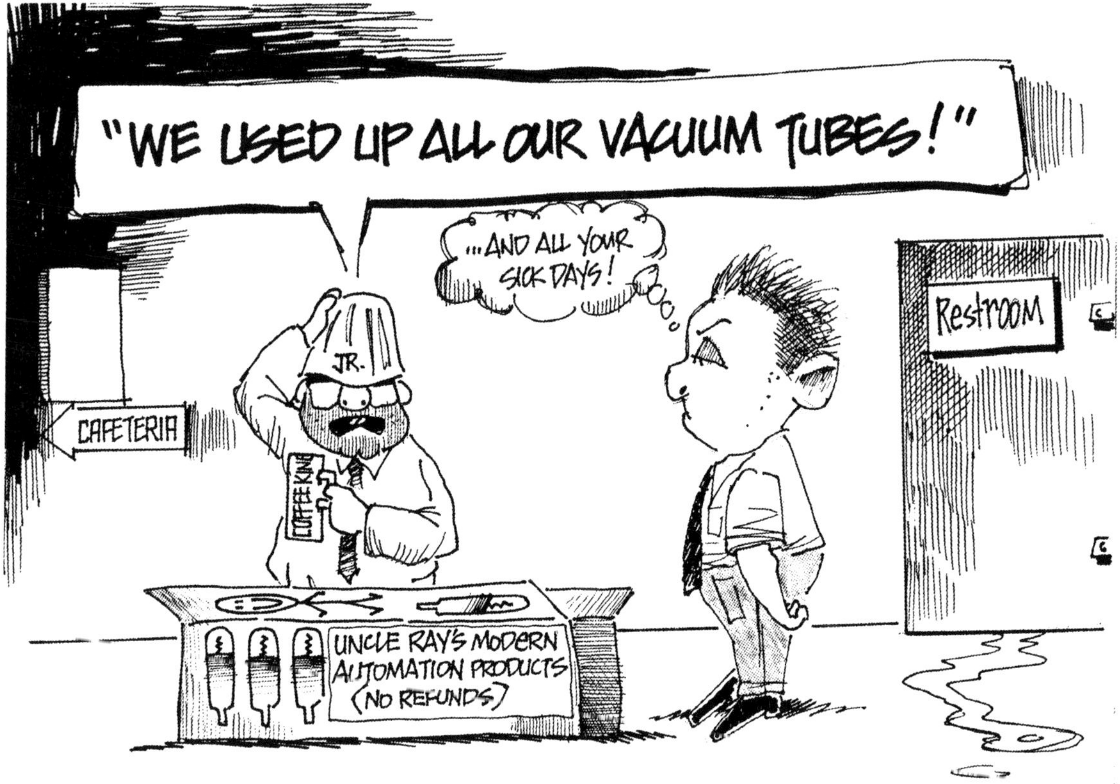
"WE USED UP ALL OUR VACUUM TUBES!"
...AND ALL YOUR SICK DAYS!
JR.
CAFETERIA
COFFEE KING
Restroom
UNCLE RAY'S MODERN AUTOMATION PRODUCTS (NO REFUNDS)

From the control room...

Uzi – "Bruce, what kind of bread is that?"
Bruce – "It's white wheat bread."

"THAT RETIRED ENGINEER THAT YOU HIRED WANTS TO SELL THE PURGE WATER FROM THE BOILERS AND COOLING TOWERS TO THE CITY!"

From the control room...

"What is that new show that just came out on BVD?" – Dub

"They should last until they fail." – Kevin H.

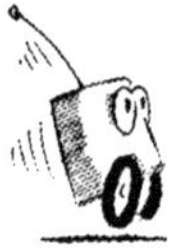

"...THE OLD DUDE IS WARNING US THAT IF WE REMOVE THE 1/4" STEEL CONTROL PANELS, THE WHOLE BUILDING WILL LIKE, FALL DOWN OR SOMETHING!"
ENGINEERS RULE
ON OFF NEUTRAL

From the control room...

"You have 20 guys out on the basketball field." – Dodie

"I feel like I'm in a wax museum. No one is moving." – Ryan

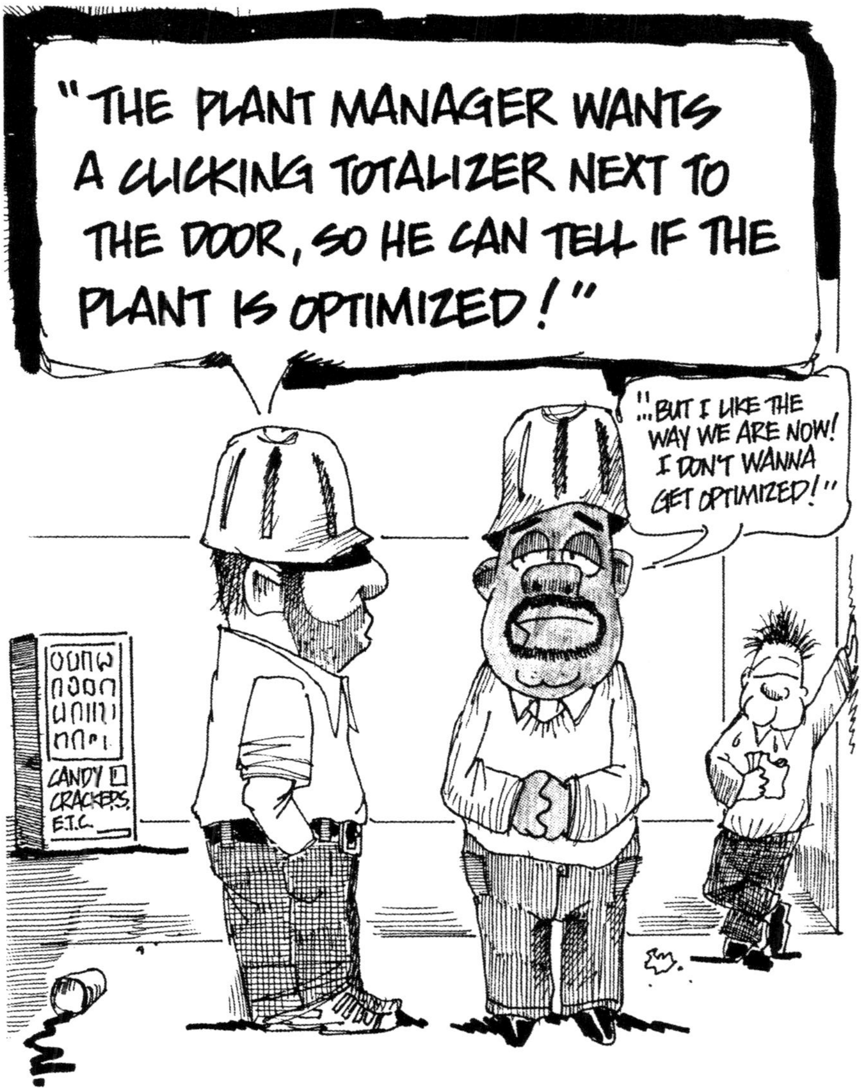
"THE PLANT MANAGER WANTS A CLICKING TOTALIZER NEXT TO THE DOOR, SO HE CAN TELL IF THE PLANT IS OPTIMIZED!"
"...BUT I LIKE THE WAY WE ARE NOW! I DON'T WANNA GET OPTIMIZED!"
CANDY
CRACKERS,
E.T.C.

From the control room...

"My wedding is the 27th of the Halloween month." – Felix

"He was a pallbearer in the wedding." – Dodie

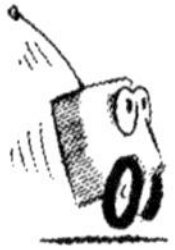

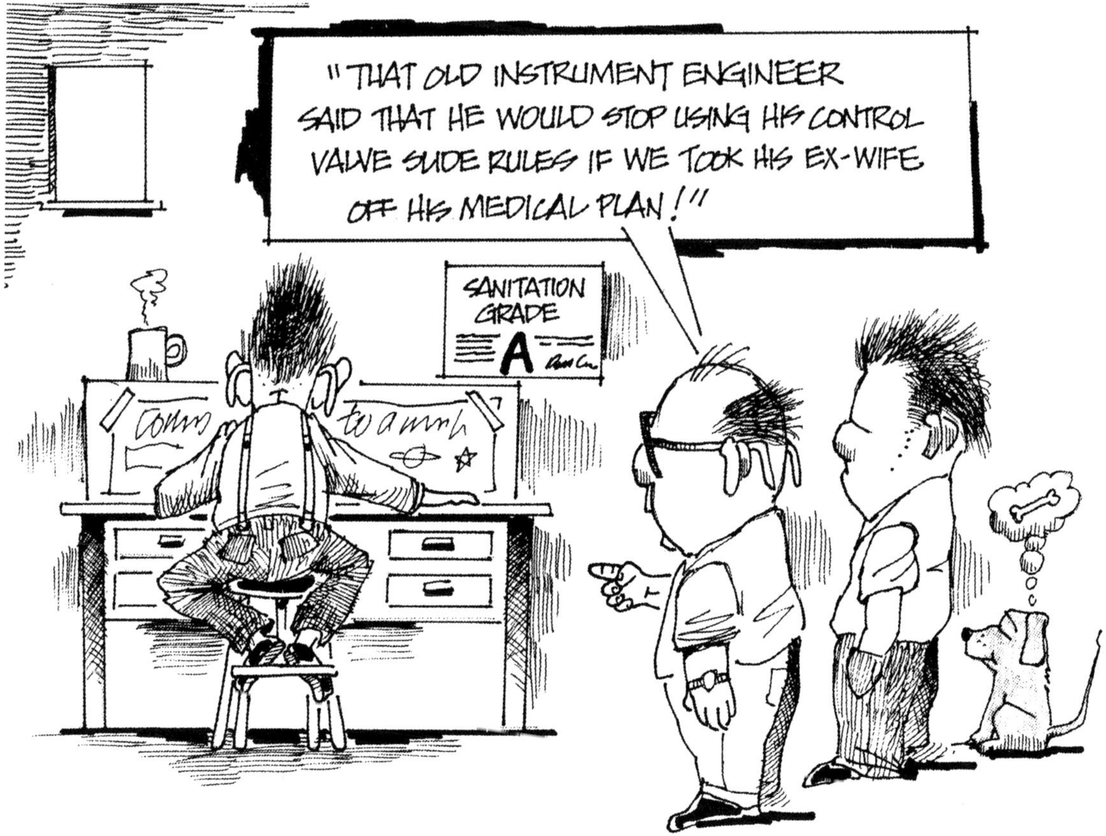
"THAT OLD INSTRUMENT ENGINEER SAID THAT HE WOULD STOP USING HIS CONTROL VALVE SLIDE RULES IF WE TOOK HIS EX-WIFE OFF HIS MEDICAL PLAN!"
SANITATION GRADE
A

From the control room...

"You could be like Larry with your earballs sunk in." – Dirck

"You can hear the air." – Quinton, during a rain storm

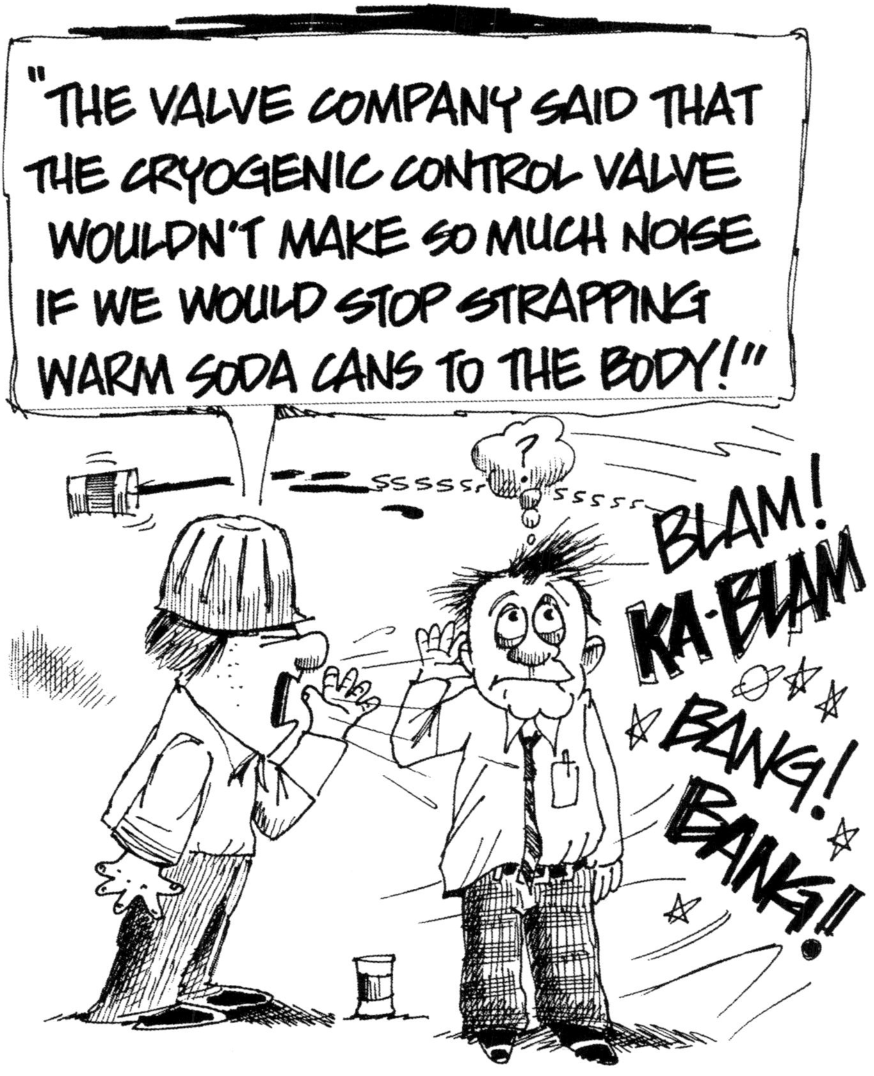
"THE VALVE COMPANY SAID THAT THE CRYOGENIC CONTROL VALVE WOULDN'T MAKE SO MUCH NOISE IF WE WOULD STOP STRAPPING WARM SODA CANS TO THE BODY!"
?
SSSSSS
SSSSS
BLAM!
KA-BLAM
BANG!
BANG!!

From the control room...

"That guy had 15 bags in his clubs." – Purdy

"Eating a grapefruit is like eating less than nothing." – Janet

"...THAT OSHA INSPECTOR SAID WE NEED TO STOP LOOSENING THE STEAM VALVE PACKING TO PREPARE FOOD... HOW DO YOU WANT YOUR EGGS?"
Rare
well done
VALVE MASTER
I ♥ GOLF

From the control room...

"I think I'll take a couple of days' vacation tomorrow." – Janet

"If you're not running so fast you are not running." – Scrap Iron

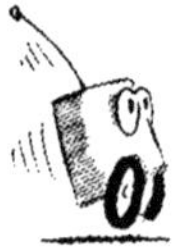

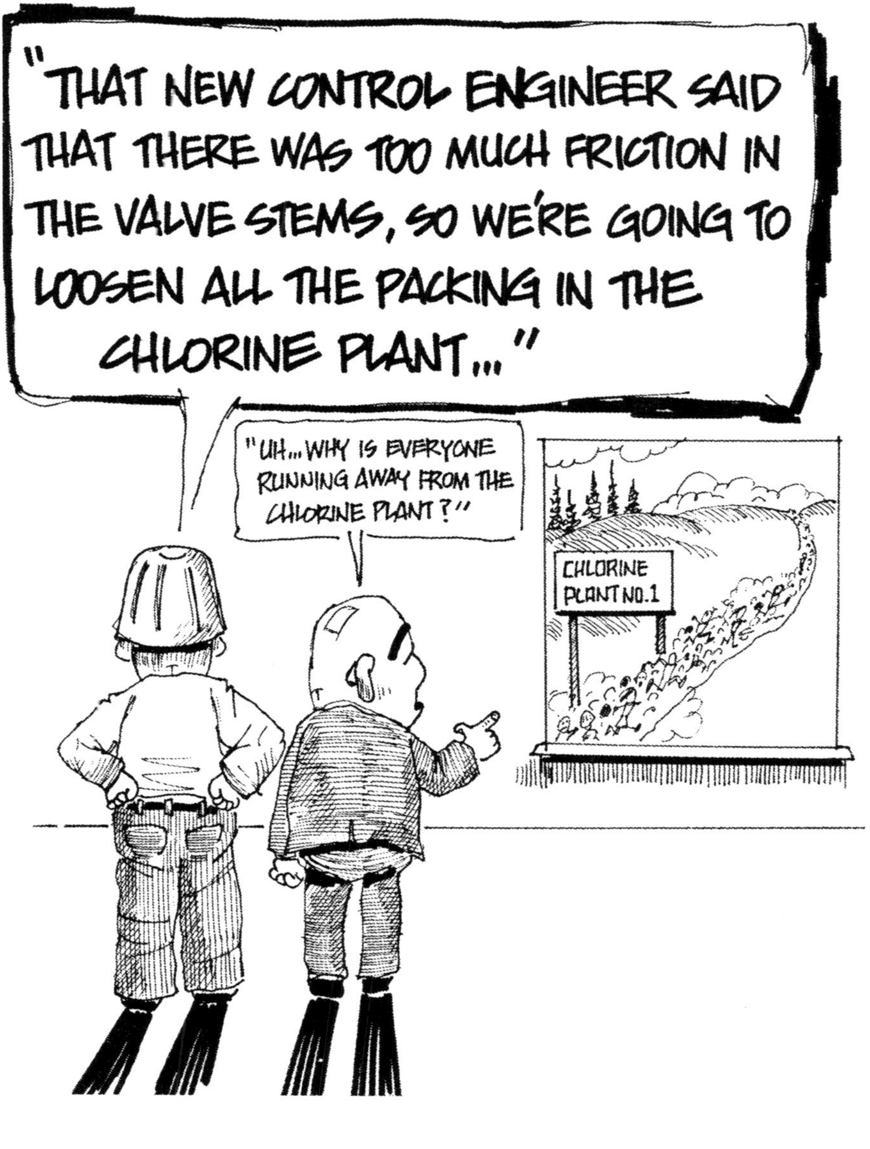
"THAT NEW CONTROL ENGINEER SAID THAT THERE WAS TOO MUCH FRICTION IN THE VALVE STEMS, SO WE'RE GOING TO LOOSEN ALL THE PACKING IN THE CHLORINE PLANT..."
"UH... WHY IS EVERYONE RUNNING AWAY FROM THE CHLORINE PLANT?"
CHLORINE PLANT NO.1

From the control room...

"Have you seen any of those gold silver dollars?" – Ron

"The law is 99/100th of the possession." – Brian

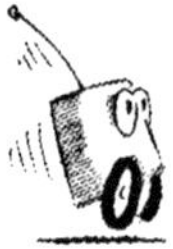

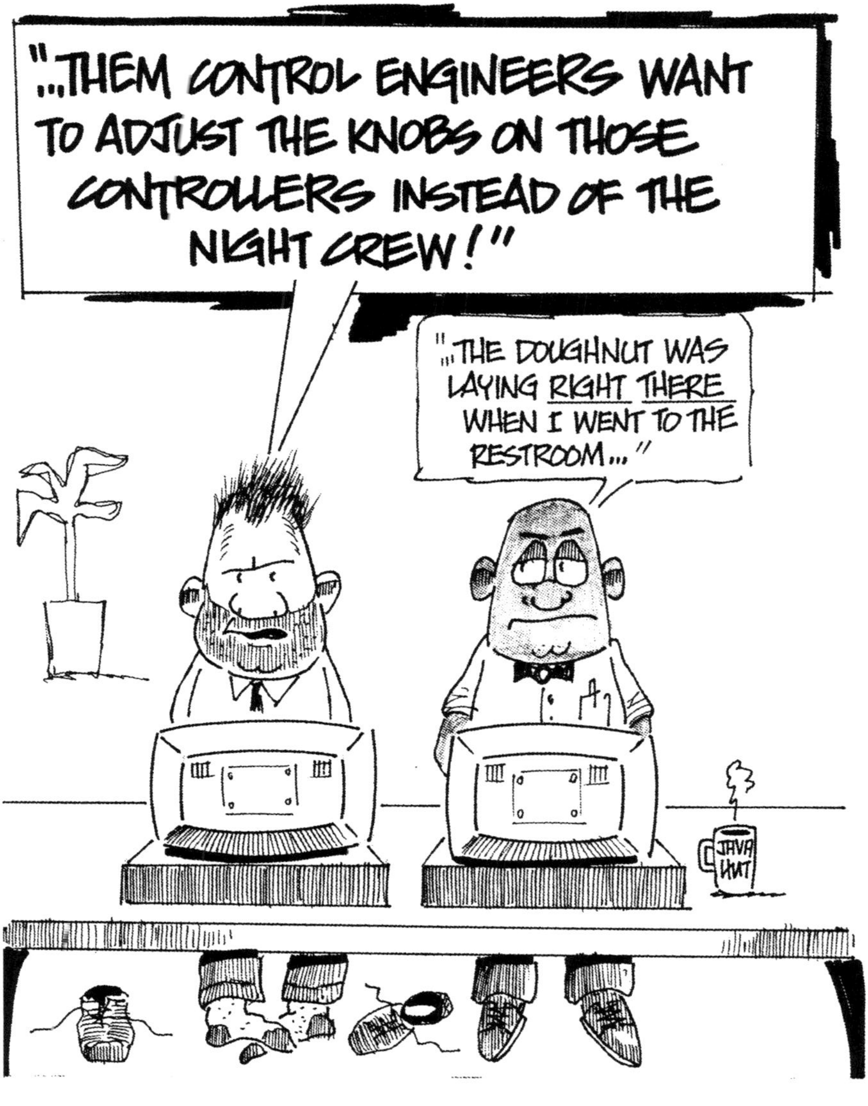
"...THEM CONTROL ENGINEERS WANT TO ADJUST THE KNOBS ON THOSE CONTROLLERS INSTEAD OF THE NIGHT CREW!"
"...THE DOUGHNUT WAS LAYING RIGHT THERE WHEN I WENT TO THE RESTROOM..."
JAVA HUT

From the control room…

"It was based on fictional facts." – Dirck

"They are trying to put a square hole into a round peg." – Dodie

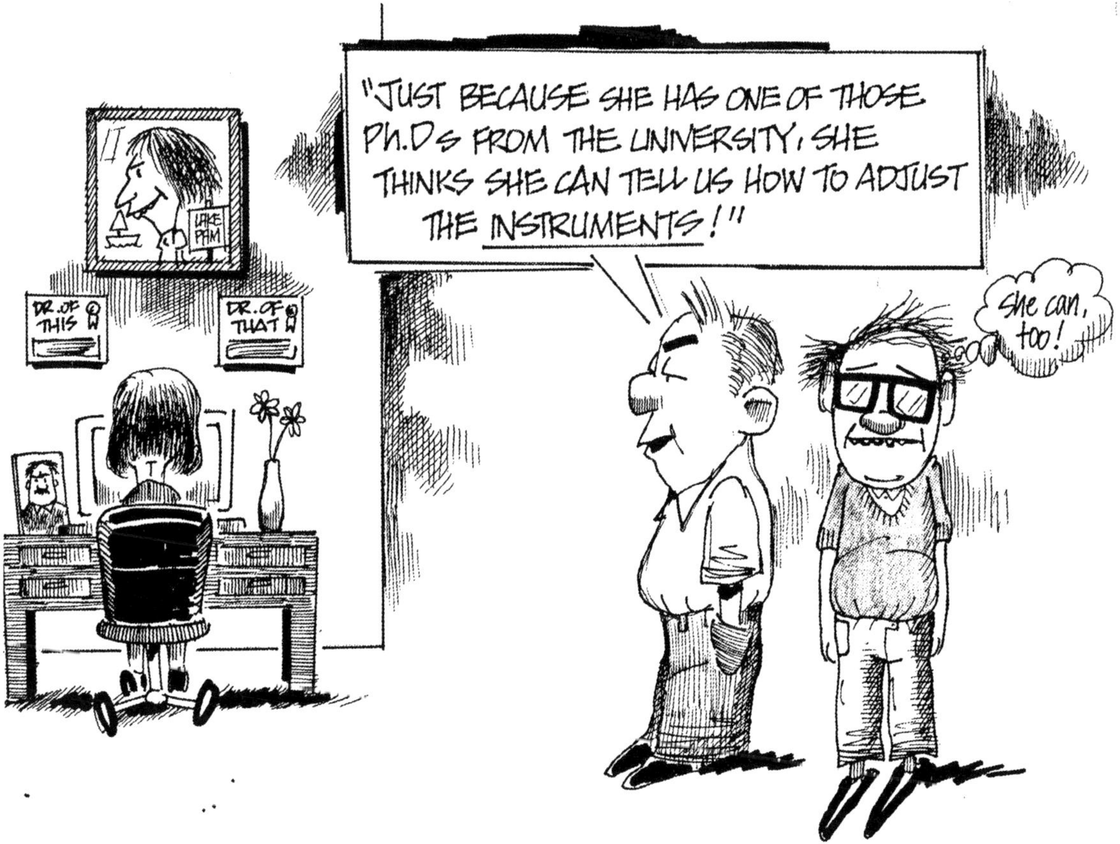
"JUST BECAUSE SHE HAS ONE OF THOSE Ph.Ds FROM THE UNIVERSITY, SHE THINKS SHE CAN TELL US HOW TO ADJUST THE INSTRUMENTS!"
She can, too!
DR. OF THIS
DR. OF THAT
LAKE PAM

From the control room…

“Where is your headache Bodene, in your back?” – Fred

“… as long as it don’t get drugged out more than a few days.” – Purdy, talking about repairs to Finisher III

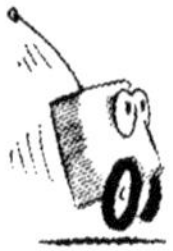

"HE'S BUILDING AN EXPLOSION PROOF 3-ELEMENT CONTROLLER FOR A POWER PLANT IN IRAQ..."
IT'S NOT QUITE READY...
HOUCHIE MAMA!

From the control room...

"He could see an elephant on a gnat's butt." – Dirck, talking about aerial photography

"A buddy of mine got his nose worked on and they didn't even use antiseptic." – Steve A.

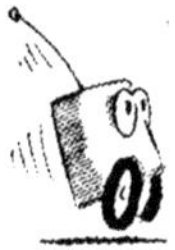

"...SO FAR, I HAVE: 2 PAGES, 788 WORDS, 3,396 CHARACTERS, (NO SPACES) 4,166 CHARACTERS WITH SPACES, 35 LINES AND 48 PARAGRAPHS. I ALWAYS WONDERED WHAT WOULD HAPPEN IF I PUSHED THAT BUTTON UNDER 'TOOLS'...."
and it's not even lunch yet...

From the control room…

"Just think Zeek, that's 18 months of sleeping in early." – Uzi

"I thought I had a hole in my lip but it was in the can." – Dodie

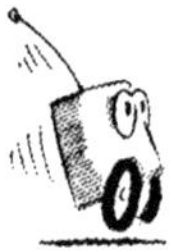

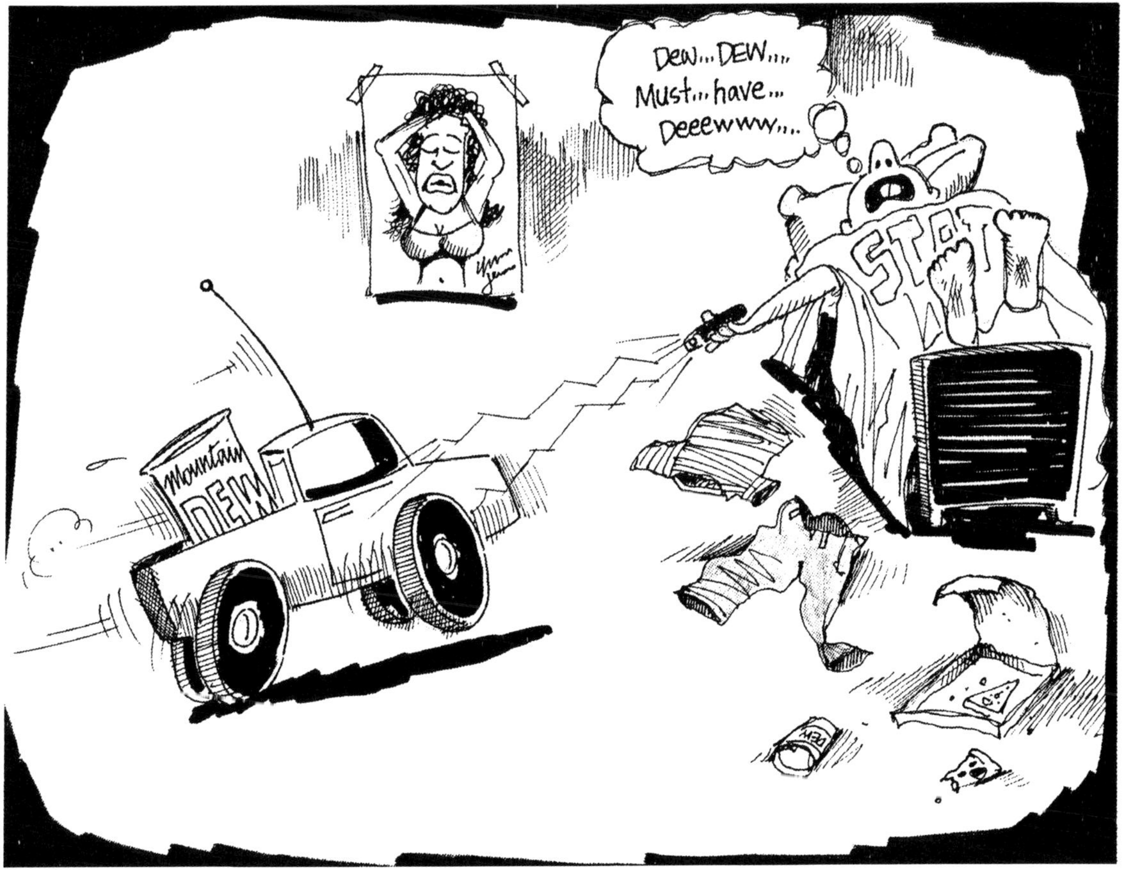
Dew... DEW....
Must... have...
Deeewww....
Mountain
DEW
STAT
DEW

From the control room...

"That's like crying wolf when there is no sheep around." – Dub

"They take the dead animals out and kill them." – Cole

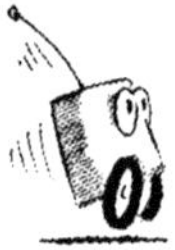

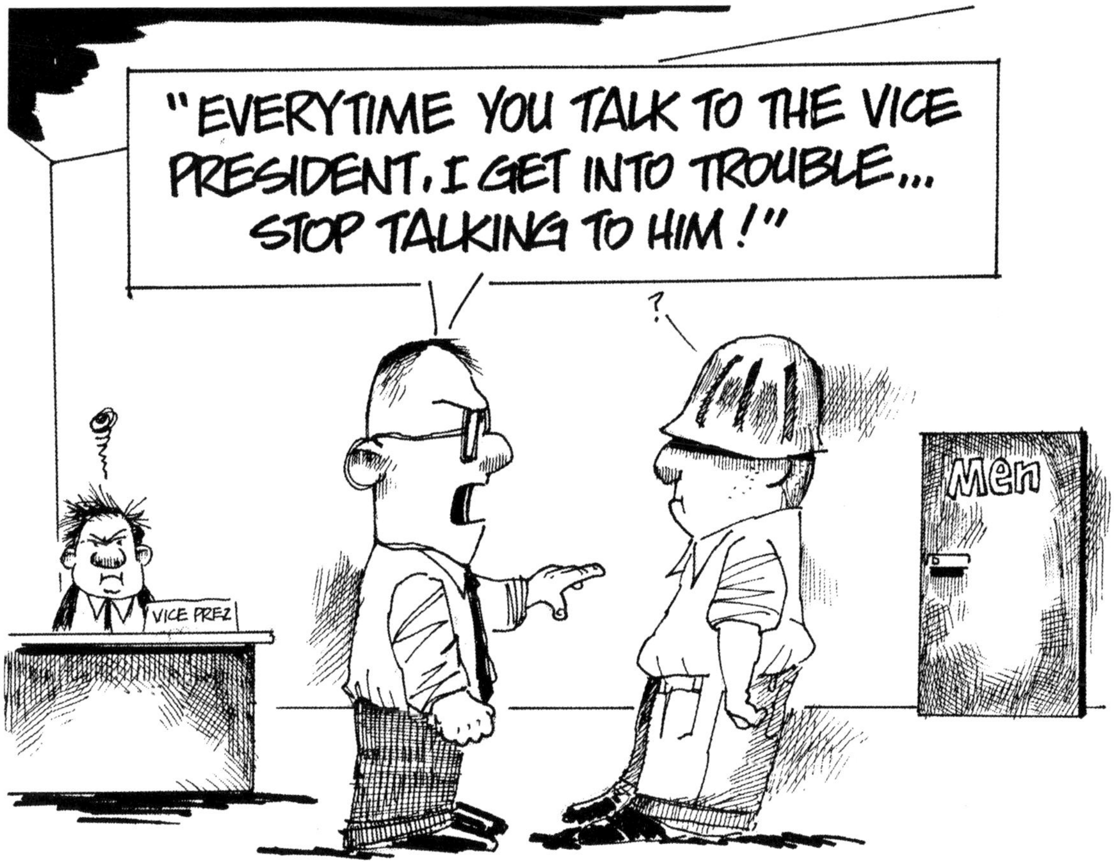
"EVERYTIME YOU TALK TO THE VICE PRESIDENT, I GET INTO TROUBLE... STOP TALKING TO HIM!"
?
VICE PREZ
Men

From the control room...

"They're turkey tenderloins. There ain't no turkey in it." – Dirck

"They drew money out of my artillery vein!" – Robbie

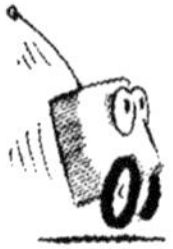

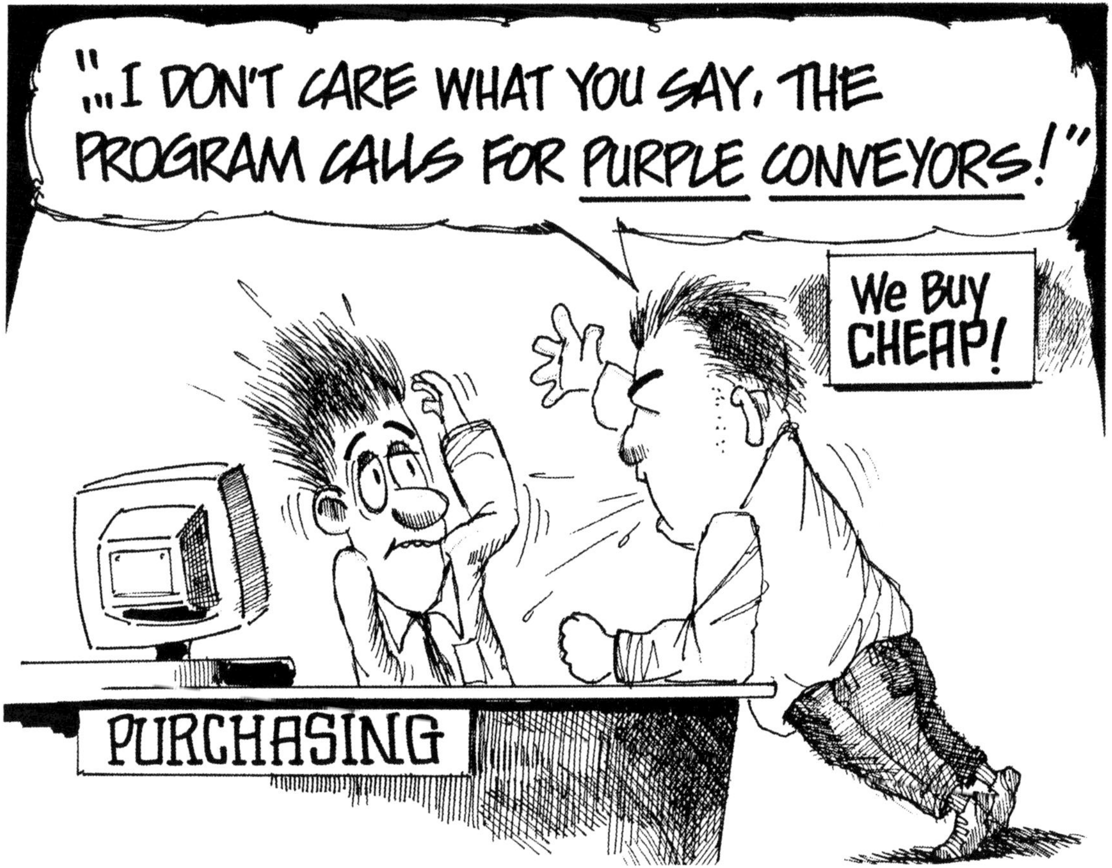
"...I DON'T CARE WHAT YOU SAY, THE PROGRAM CALLS FOR PURPLE CONVEYORS!"
We Buy CHEAP!
PURCHASING

From the control room...

"I went from smoking Lucky Strikes to smoking methanol." – Dub

"The stuff coming over the weir is dark black." – Uzi

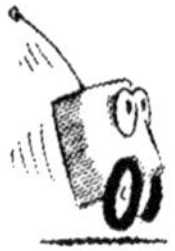

HELL
...NO! HE IS!
...NO! HE IS!
"...WE HAVE A HOT OPPORTUNITY! WHICH ONE OF YOU IS THE AIR CONDITIONING SPECIALIST?"
OPEN 24-7
THE DEVIL

From the control room...

"Those are the top of the art." – Dub

"They have a ten foot gate all around the place." – Purdy

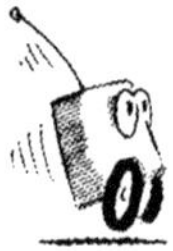

I AM!
NO! I AM!
..."GOOD GRACIOUS, GENTLEMEN! SORRY FOR THE MIX UP! WE MEANT TO KEEP AN AIR CONDITIONING SPECIALIST... SO... WHICH ONE OF YOU IS INTO AIR?"
HEAVEN
CLOSED SUNDAYS
NO SMOKING
ST. PETER
YOGURT

From the control room...

"What time does the noon group tee off?" – Brian W.

"There is no white water rapiding in North Carolina." – Uzi

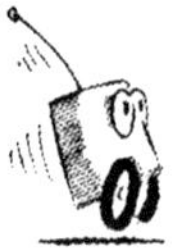

"...NOTHING LIKE A CAREER IN AUTOMATION TO END UP WITH A SWIMMING POOL EXACTLY AT THE RIGHT TEMPERATURE AND PH LEVEL TO BOOT... AAAHH...."
NO RUNNING

Certifiable Top Ten Lists

After the core dump of "Top Ten Lists" that went into the book *Dispersing Heat by Conviction* and startups were only a distant memory, we briefly considered maybe the best lists were behind us. The preparation of a monthly column in *Control* magazine and the contributions of other graduates of the Monsanto family, notably Bob Cameron, Glenn Metz, and Mark Sowell, have enabled us to carry on the tradition. The following is a collection of Top Ten Lists from our "Control Talk" column since its inception in 2002.

Top Ten Signs a Startup Has Gone Wrong

10) Product tanks are empty.
9) Waste tanks are full.
8) The startup team and DCS have negative free time.
7) The entire design team is in the control room.
6) The whole research team is the control room.
5) The managers have left the control room.
4) The consoles are lit up like pin ball machines.
3) Operators keel over from doughnut overdose.
2) A collection is taken for the startup budget.
1) The order for souvenir baseball caps is cancelled.

Top Ten Ways to Maintain Your Sanity on the Job

10) Insist on being addressed by your superhero's name, "The Great Automator."
9) Put mosquito netting around your work area and play tropical sounds all day.
8) As often as possible, skip rather than walk.
7) Finish all your sentences with "in accordance with the prophecy."
6) Put decaf in the coffee maker for 3 weeks. Once everyone has gotten over their caffeine addictions, switch to espresso.
5) Put your garbage can on your desk and label it "In".
4) Every time someone asks you to do something, ask if they want fries with that.

3) Page yourself over the intercom. Don't disguise your voice.
2) At lunchtime, sit in your parked car with sunglasses on and point a hair dryer at passing cars. See if they slow down.
1) Wear headphones and bring your favorite comic book to departmental meetings.

Top Ten Signs of a Valve Problem

10) The pipe fitters are complaining about trying to fit a 1-inch valve into a 10-inch pipe.
9) You bought the valve suppliers' "monthly special."
8) A butterfly disc won't open because the lined pipe ID is smaller than the disc OD.
7) The maintenance department personally put the valve on your desk.
6) A red slide ruler was used to size a green valve.
5) Your latest valve catalog is dated 1976.
4) The maintenance department said they don't want a double seat "A" body.
3) The valve was specified to have 0% leakage for all conditions including all signals.
2) The fluid field in the sizing program was left as water.
1) The valve is bigger than the pipe.

Top Ten Holiday Gifts for Control Engineers

10) A book titled, "Project Management Made Easy" or "How to Rewire the Instruments on the 3rd Shift".
9) Short-life batteries for your pager that will conveniently die when you get home.
8) "New and improved" products that didn't break the features of the previous version that were critical to your operation.
7) An automated voice recognition system and translator for the help line.
6) A fully automated brewery in your basement.
5) A remote control for your mother-in-law with a mute button.
4) Tee shirt with "Control Freak in the Room" on it.
3) A swimming pool with Model Predictive Control.

2) Life size photos of *Models Unleashed*.
1) The "Swimsuit Issue" of Control.

Top Twenty Signs Your Son Knows More about Computers than You

20) You are wired and your son is wireless.
19) Your son is working and living in Silicon Valley with two more computer jocks whose girlfriends work for software companies. You call to ask why your computer won't do anything. After a long conference, the five of them tell you to shut it off.
18) After one week with the new computer, your 12-year old is rewriting the game programs because they are too easy.
17) Your 15-year old is rewriting the control valve sizing programs because the local vendor doesn't understand the manufacturer's version.
16) Your son gets a higher starting salary than you make after 30 years on the job.
15) Your son refuses to work on your system because it is too primitive.
14) Y0ur k1d 5p34k5 1n 1337 4nd y0u c4n7 und3r574nd 17.
13) You use your computer for playing games and he uses it for writing games.
12) You use your computer for watching videos and he uses it for making videos.
11) You find a website that's really cool and later learn that your son created it.
10) You still use Windows 95 (or 98, or 3.1, or NT) and he is telling you about the expected advancements in Longhorn.
9) Your kid is using operating systems that you haven't even heard of.
8) Your kid has more computers than the rest of the family combined.
7) You use a soldering iron to repair old stereos and he uses it to repair old motherboards.
6) You don't know how many megs are in your son's terrabyte RAID array.
5) Your son tells you he wants to major in EE and you think he means Elementary Education.

4) You think that when your son is doing peer-to-peer he's talking with friends.
3) Your son asks for Ethernet and when you show him a bundle of 10-Base2 coax, you're promptly informed that it's about 15 years out of date.
2) Your kid is studying networking and you wonder why he needs a course on people.
1) Your son can turn off the parental controls faster than you can set them up.

Top Ten Signs of an Advanced Control Addiction

10) You try to use Neural Networks to predict the behavior of your children.
9) You attempt to use fuzzy logic to explain your last performance review.
8) You use so much feed forward, you eat before you are hungry.
7) You propose Model Predictive Control for the "Miss USA" contest.
6) You develop performance monitoring indices for your spouse.
5) You implement adaptive control on your stock portfolio.
4) You carry wallet photos of Auto Tuner trend results.
3) You apply dead time compensation by drinking before you go to a party.
2) You recommend a survivor show where consultants are placed in a stressed out old pneumatic plant with no staff or budget and are asked to add advanced control to increase plant efficiency.
1) Your wife has to lure you to bed by offering "expert options" for advanced control.

Top Ten Ways to Impress Your Management with the Trends of a Control System

10) Make large set point changes that will zip past valve dead band and nonlinearities.
9) Change the set point to operate on the flat part of the titration curve.

8) Select the tray with minimum process sensitivity for column temperature control.
7) Pick periods when the unit was down.
6) Decrease the time span so that just a couple data points are trended.
5) Increase the reporting interval so that just a couple data points are trended.
4) Use really thick line sizes.
3) Add huge signal filters.
2) Increase the process variable scale span so it is at least ten times the region of interest.
1) Increase the historian's data compression so that most changes are screened out as insignificant.

Top Ten Signs that a Batch Has Gone too Long

10) You have eaten so many doughnuts you look like the "Pillsbury Dough Boy."
9) The operators start showing you their favorite pictures of guns, cousins, and livestock.
8) Your CEO is on the phone holding for you.
7) The contract engineers are going online to buy SUVs with the overtime pay.
6) You suspect the contract firm's last job was the Mars Rover.
5) The phase of the batch appears correlated to the phase of the moon.
4) The displays appear to be in suspended animation.
3) The batch executive is referred to as "Hal".
2) Jupiter's moons are starting to look good as your motto becomes "anywhere but here."
1) Your company changes names several times before the batch is done.

My Top Ten Reasons for Becoming a Control Engineer

10) To meet girls. Unfortunately there were none in my class or on my job. I guess I was ahead of my time.
9) To impress girls. Boy was I wrong again. Have you ever tried to describe what you do as a control engineer in a bar?

8) I mistook all those personal ads asking for a "control freak" in the *Riverfront Times* as a career opportunity.
7) It offered a chance to carry photos of my favorite pH electrodes in my wallet. Now I can carry CDs.
6) I thought the control room would be like the bridge of the Starship Enterprise. I didn't find any anti-matter drives but I did get to experience strange forces in project meetings, parallel universes in plants, and spin during performance reviews.
5) I didn't know the job would be teaching pipe fitters how to calibrate and install pneumatic instruments. I spent my formative years as a site construction engineer in West Virginia where the plant instrument technicians didn't show up until after the units had started up and were running smoothly. Somehow I missed the courses on how to swap knives and accurately spit tobacco juice in a can or on a project manager's shoes.
4) An opportunity to learn "down home" cooking. Some of the best dinners I have had have been prepared in the kitchens of our Cajun control rooms.
3) The flashing lights and sounds of big time action. As technologies advanced, so did the number of alarms. Now each point comes with 5 configurable alarms.
2) The chance to get on the cover of *Control* magazine. My dream came true last December.
1) It was the only job offer. Majoring in engineering physics, I should have been suspicious when all of the other students were nuclear submarine officers.

Top Ten Signs Your Company Has Downsized Too Much

10) Consultations with yourself are the norm.
9) It's impossible to create a normal distribution of employee performances.
8) None of your co-workers are employed by your company.
7) The signatures on your "retirement" plant photo barely fill 25% of the border.
6) The common reply to task requests is "yeah, whatever."

5) The new corporate phone directory is distributed on a 3 x 5 note card.
4) The human resources department is a bulletin board in the lobby.
3) You call in, request to be transferred to your boss's secretary and get forwarded to yourself.
2) The "auditorium" for large corporate meetings is now the cubicle next to you.
1) Retirement "packages" are a box with the personal belongings from your desk.

Top Ten Email Notices Indicating Your Company Is Going Offshore

10) We have assigned you a bodyguard for your next trip.
9) Do you know any Arabic, Chinese, Russian, or Sanskrit?
8) Beards and turbans will be standard issue.
7) We have signed a corporate agreement with Hertz "Rent a Camel."
6) Fraternization with monkeys is strongly discouraged.
5) Can your job be done remotely?
4) Is there a closet big enough for the engineering department's home office?
3) We have an offer you cannot refuse.
2) The CEO has an important telecast for developing "World Class" engineering.
1) The name of your department is changed to "Virtual Reality."

Top Twenty Signs a Spin-off May Be Spinning toward Bankruptcy

20) As a parting gift your new company gets 35% of the assets and 90% of the debt.
19) You stop receiving copies of the free trade journals.
18) The bathrooms suddenly have toilet paper dispensers that accept quarters.
17) The "prior convictions" question is omitted from the company's employment application.
16) Human Resources laugh when you ask to sign up for the stock-based 401K plan.
15) There's a tollbooth at the entrance to the company parking lot.

14) You see an ad in the paper for a real estate open house at your company's address.
13) There have been an unusually high number of visitors with tape measures.
12) You find paper airplanes made from company stock certificates.
11) The only cost reduction plan implemented is head count reduction.
10) The CEO gets a huge bonus for losing less money than the budgeted loss. (Oh heck, give him one even if it's more just to ensure the captain goes down with the ship.)
9) The new medical benefits plan consists of a pair of dice.
8) You try to pay your suppliers with Monopoly money.
7) The executives are replaced with lawyers.
6) Domino's won't accept your check.
5) You have three phone messages from your financial planner.
4) You see your boss at the mall buying a new "interview" suit.
3) The evening shift Wal-Mart greeter vacancy was filled by one of your colleagues.
2) There are orange survey flags along the plant property lines.
1) Zeb the janitor is your DCS supplier's representative at your company's Vendor Alliance meeting.

Top Ten Signs You Are Ready for a Hawaiian Vacation

10) You give your boss the "hang loose" hand gesture.
9) You daydream about hula dancers in hardhats.
8) Your cubicle has a mosquito net with tropical sounds.
7) You bring a kayak to the company's waste pond.
6) You ask, "Where is the company's pupa stand?"
5) You tell your secretary she is wearing a nice muumuu.
4) You play a ukulele in your office.
3) You show up to a meeting in a Hawaiian shirt, shorts and thongs.
2) You start answering your phone saying, "Aloha."
1) You wear a snorkeling mask instead of glasses.

Top Ten Handy Uses for Large Case Circular Chart Recorders

10) Big bird feeders, or in my case, squirrel feeders.
9) Flower planters – you can pass them off as antiques.
8) Portrait frames for your long lost relatives.
7) Halloween masks for aspiring engineers.
6) Piggy banks so you can get the money back out.
5) Storage for your memorabilia such as valve sizing slide rulers.
4) Hi-tech picnic games – stand back 20 feet and toss your CDs.
3) Great nautical-themed lockers with those portholes.
2) Bourdon tubes make nice fishing lures - just add treble hooks.
1) Toilet seats for "Outhouses" – what an appropriate end.

Top Ten Signs you are an Endangered Species

10) You find conversations with yourself most productive.
9) Your discipline is no longer on any organization charts.
8) Your technical society changes its name.
7) Vendors go into a feeding frenzy when you show up at a society meeting.
6) Your youngest colleague is eligible for senior discounts.
5) The "Users Group" meeting is held at a retirement resort.
4) A natural history museum asks for your job paraphernalia.
3) The local zoo requests to put you on exhibit.
2) You become the "Poster Child" for the Sierra Club.
1) Your mating habits are featured on the "Discovery Channel."

The ranks of automation professionals have become perilously thin so we conclude this book with a vain attempt to convince you not to crowd the retirement resorts. This new top ten list was concocted by Glenn Mertz prior to beer o'clock. If you want to improve them, just add beer.

Top Ten Reasons to Not Retire

10) You would really miss those 2-hour staff meetings every Friday morning.
9) You would have trouble finding how to use those hours you spend on your commute.
8) You can't bear the thought of not starting out your day with a delicious office cup of instant coffee under the flickering florescent lights with the crackle of office sounds.
7) You would have too much trouble finding reasons to use PowerPoint.
6) You could not live without an annual performance review.
5) Your fitness would be compromised by missing the strength training of carrying your laptop all over the world.
4) It is too difficult to figure out what you would do with your office wardrobe.
3) You are afraid of withdrawal symptoms if you can't spend 8 hours a day on the phone and doing email.
2) You don't like the idea of having 52 weeks of vacation every year.
1) The company canteen vending machines have the best hot dogs in town.